The
Physical
Properties
of
Organic
Monolayers

The
Physical
Properties
of
Organic
Monolayers

Mitsumasa Iwamoto
Tokyo Institute of Technology

Wu Chen-Xu
Tohoku University

World Scientific
Singapore • New Jersey • London • Hong Kong

Published by

World Scientific Publishing Co. Pte. Ltd.

P O Box 128, Farrer Road, Singapore 912805

USA office: Suite 1B, 1060 Main Street, River Edge, NJ 07661

UK office: 57 Shelton Street, Covent Garden, London WC2H 9HE

British Library Cataloguing-in-Publication Data
A catalogue record for this book is available from the British Library.

THE PHYSICAL PROPERTIES OF ORGANIC MONOLAYERS

ISBN 981-02-4482-7

Printed in Singapore by Uto-Print

PREFACE

Organic monolayers on material surfaces exhibit various behaviors that are interesting to researchers from viewpoints of physics, chemistry and biology. Since the discovery of the technique for the formation of monomolecular films on water surface by I. Langmuir (1881-1957), along with the development of methodologies, innumerous experimental works have been carried out in this field and many insights into ultrathin organic films have been gained. Until now, mechanisms of the specific properties of two-dimensional monolayer systems, though not sufficient, have been elucidated step by step. In recent years one sees a considerable growth of interest in ultrathin organic films because of its potential applications in nanoelectronic devices. The investigations on their dielectrics, optical reaction, electroluminicence, electronic properties, and electrochemical reaction can be found in various literatures. However, the properties of specific two-dimensional systems have not been fully discussed yet from the view points of physics and electronics. This is the main subject of this book.

For ultrathin organic films, the introduction of surfaces breaks the space-inversion symmetry of conventional bulk materials such as nematic liquid crystals. As has been supported by various kinds of experiments such as Second Harmonic Generation (SHG) and Maxwell Displacement Current (MDC), this gives the characteristic physical properties of organic monolayers that, in many cases, are quite different from those of their bulk counterparts. Looking at R&D in this field, one will find that the two most important keywords for ultrathin organic films are "polar" and "interfacial", reflecting the characteristic symmetry-breaking effect and the thickness effect respectively.

Whenever a material system such as crystals and liquid crystals is studied, scientists of statistical physics prefer to start from the characteristic structures of the system using order parameters. By introducing order parameters to express the non-centrosymmetric monomolecular systems, it is possible to analyze the specific physical properties of organic monolayers in terms of structural parameters. This provides an approach to bridge the physical properties such as dielectric or electronic with the chemical texture of the system, which will be of great help for constructing the theoretical basis for the development of nanoelectronics using organic materials. Considering this, the authors entitled the book as "THE PHYSICAL PROPERTIES OF ORGANIC MONOLAYERS".

This book is intended to give a fundamental physical picture of various phenomena occurring in organic monolayers, covering dielectric, elastic, and electronic properties. Our treatments, focusing on the explanations of the structure dependence (dipole moment) and thickness dependence (interfacial effect) for various phenomena occurring in organic monolayers, are restricted within mean field theory. Those who want to dig deeper may resort to fluctuation perturbation theory or renormalization group treatment. The target readers include postgraduate and researchers in the fields of organic materials chemistry, nanoelectronics, dielectrics, low-dimensional physics, and interfacial physics, materials science etc.

This book could not have been completed without the substantial support by Prof. Z. C. Ou-Yang, Institute of Theoretical Physics, Academia Sinica, People's Republic of China. Some chapters of this book are based on the collaborative work between Prof. Ou-Yang and the authors. M.I. would also like to thank Dr. T. Hino, professor emeritus of Tokyo Institute of Technology, Japan, for his priceless advice and encouragement over long and arduous years of research activities. Gratitude is also given to visiting scientists and students who were once engaged in their researches and studies in Iwamoto's Laboratory in Tokyo Institute of Technology. We will be very appreciated to hear comments, critiques, suggestions from any readers for potential revisions and additions.

Department of Physical Electronics
Tokyo Institute of Technology, Japan   Mitsumasa Iwamoto

Institute of Fluid Science
Tohoku University, Japan   Chen-Xu Wu

July, 2000

CONTENTS

CHAPTER 1

INTRODUCTION

Depositing monomolecular assemblies on various substrates, now termed Langmuir-Blogett (LB) technique, is an experimental field of science offering a lot of powerful tools to study various problems in physics, chemistry, electronics, biology, and others. [1, 2, 3] The preparation of Langmuir films, monomolecular assemblies on a water surface, is carried out by depositing a small amount of an amphiphile solution in a volatile solvent onto a clean water surface and then a monolayer film is formed as the solution spreads spontaneously and the solvent evaporates. [4] Once the LB technique was first discovered in 1920s, it proved to be an easy and reliable method in studying the physicochemical properties of ultra-thin films. This measuring deposition approach with the help of various newly developed measurement instruments raised the boom of the investigation on ultrathin films in the past seventy years. The richness in the variety of organic materials suitable for LB deposition offers enormous scope for those who are interested in the field of material science. In association with other measuring techniques, the molecular dynamics of a large variety of materials can be probed by simply depositing the ultra-thin film of the same materials on a water surface or on a material surface.

Until now, the vigorous developments of microelectronics and optoelectronics have relied largely on inorganic materials such as silicon and lithium niobate in single crystal form. [1] However, as the perceived limitations of these materials restrict the realization of more complex system designs and as a result will foreseeably paralyze their further development, the rapid progress in LB films in the past decade exhibits their great importance and allows the potential applications of organic device in these fields. The appearance of several kinds of state-of-the-art measuring instruments, such as Atomic Force Microscopy (AFM), Scanning Tunneling Microscopy (STM), [5] Second-Harmonic Generation (SHG), Surface Potential Measurement (SPM) in the past several decades, has accelerated to some extent the progress and strengthens the belief in the development of organic films as new generation devices. Some enthusiasts even believe that it is absolutely possible in the near future to replace the conventional inorganic devices with organic devices in certain areas such as the processing, the transmission or the storage of information.

Physical properties always come from structural features. There is no exclusion so far concerning the correspondence between the material structures and the material properties for crystalline solids as well as for liquid crystals. The crystalline solids exhibit strict regulations of physical properties due to nothing more than their periodical lattice structures. The system, as a whole, is isotropic, but at the same time reveals some anisotropic properties on a local scale. One example is the liquid crystals, which is an ordered system with a lower symmetry than crystalline solids.

1

The liquid crystals show striking anisotropy, a most important feature widely used in their applications. Because of the structural difference between organic monolayers and liquid crystals arising from their *polarity*, as will be seen later in the following, it is natural that their dynamics should be different from that of liquid crystals. This is the theoretical basis for the following chapters and the major reason for writing this book.

1.1 Monolayer Structure and Monolayer Properties

A classic monolayer-forming material such as fatty acid, e.g., stearic acid and arachidic acid, has two parts in the molecule: a *hydrophilic* head group, which is easily soluble in water, and a long alkyl chain, which provides a *hydrophobic* tail. Figure 1.1 shows the structure of fatty acid on a water surface. Such a structural feature completely

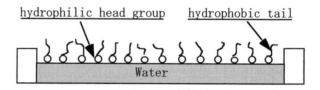

Figure 1.1 Monolayer of fatty acid on a water surface.

excludes the possibility of the anti-ferroelectric orientation of dipolar molecules in Langmuir films. In most cases, monolayer films on a material substrate have the similar tendency of para-ferroelectric orientation due to the interfacial interactions between the monolayer and the substrate. This structural feature reveals that monolayer can be regarded as a system with C_∞ symmetry, which is low compared with $D_{\infty h}$ of nematic liquid crystals. Such a symmetry enables one to regard the monolayer system as a kind of "polar LCs" and reminds one of the famous Brownian rotational motion, which proved to be a suitable model for monolayer films. [6] Basically, the dynamics of the constituent molecules in monolayer films can be decomposed into two kinds of motions, the planar positional motion and the orientational motion of the constituent molecules. [7] When the electric properties of the monolayer films, which largely depend on the orientation of molecules, are mainly concerned, a reasonable approximation can be made that the molecules are identical rodlike dipoles. The experimental results of some organic materials such as 4-cyano-4'-n-alkyl-biphenyl (nCB) show a good agreement with such a model.

1.2 Surface Pressure/Area Isotherm

The first significant advance marking the cornerstone of LB technique and further triggering interest in this field came with the successful observation of the surface pressure-area plateau, of which the physical background is deeply related to the molecular structure of monolayer films. Its measurement is usually carried out by compressing the film at a constant rate while continuously monitoring the surface pressure at a constant temperature. The pressure-area relation, known as the surface pressure/area isotherm, gives the most important indicator of the monolayer properties of a material in the past several decades of its development. A typical surface

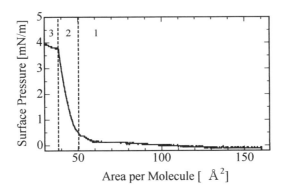

Figure 1.2 Surface pressure/area isotherm of 4-cyano-4'-5-alkyl-biphenyl.

pressure/area isotherm, as shown in Fig. 1.2, reveals that at large area, the surface pressure is very close to zero and the interactions between molecules are supposed to be negligibly small. The monolayer in this region is considered to be in an *air* (or *gas*) state (region 1). As the monolayer is further compressed, the surface pressure will increase and the monolayer change to *liquid* state (region 2) and finally to *solid* state (region 3). Many phenomena have been clarified by such a pressure/area plot, though the physical mechanism of some phenomena still remained uncertain for several years just by such an approach.

1.3 Maxwell Displacement Current Measurement Technique

The Maxwell displacement current (MDC) measurement technique, which was developed by Iwamoto Lab in 1980s, provides an indirect experimental approach to detect the molecular behaviors in monolayer films through measuring the short-circuited

MDC. [8] Suppose a dipole moment on a water surface or on a material substrate is subjected to an external stimulus (Fig. 1.3), such as light, pressure, heat or electric field, the dipole moment will change its orientation as a result of the external stimulus. Consequently, charges will be induced on the upper and lower electrodes and a displacement current will flow across the electrometer if connected between the two electrodes. Put the other way around, we can detect the molecular motion in a monolayer film by measuring the Maxwell displacement current. What we have to do is quite simple: stimulate the monolayer film and measure its short-circuited Maxwell displacement current. As will be explained in detail later in Chapter 3, the

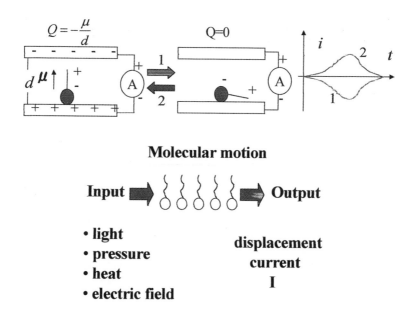

Figure 1.3 Mechanism of Maxwell displacement current measurement technique.

measurement is performed by recording the MDC signals by compressing a monolayer film at a constant rate while continuously monitoring the surface pressure-molecular area isotherm. This is attributed to the orientational change of the dipole moments in monolayer films. Its merit lies in that the two electrodes do not have to come into contact with monolayer films, thereby protecting the structure of the films from being destroyed during the measuring process. It has been proven that the molecular dynamics in monolayer films under external stimulation can be investigated through examining MDC as an output. MDC tells much about the collective properties of molecular motion. The various experimental results achieved so far by such an approach show useful evidence that the molecular behaviors in quasi two-dimensional (2D) monolayer films are different from those in bulk state. [9] This is possibly due to the interfacial interaction between the monolayer films and the material surface, which is specific for monolayers, and the intermolecular interaction within the monolayer films. For monolayer films under compression, MDC is a more feasible measuring technique, as the monolayer structure varies in the monolayer compression process, which otherwise, can be very difficult. It is interesting here to note that one can also generate MDC by vibrating the suspended upper electrode since the induced charges depend on the spacing between the upper and lower electrodes. However, this is not essential. Using fixed electrode arrangement, one can gain many useful information on the collective properties of monolayeres by stimulating the monolayers.

On the theoretical side, many attempts have been made to interpret the displacement current flowing across the organic monolayers. [10] The nematic order parameter, which is represented as an average of the second Legendre polynomial or known as Meier-Saupe order [see Eq. (1.1)], has long been used in bulk liquid crystals. [11] However, the Meier-Saupe order is not suitable for describing the polar orientational order of monolayers on a water surface, [12] as it gives no detailed information on the polar orientational order of monolayers at the air-water interfaces. In 1994, Sugimura *et al.*, using the order parameter represented by an average of first Legendre polynomial, considered the amphiphile monolayer as half a membrane and thermostatistically studied the orientational order of monopolar-molecule monolayers at the air-liquid interface without delving into the chemical details of the constituent molecules. [6] They then discussed the phase transition of 4-cyano-4'-5-alkyl-biphenyl (5CB) molecules from the planar isotropic surface alignment phase to the polar one in the range of low surface pressure during monolayer compression. In their study, the constituent polar molecules of monolayers were assumed to be rodlike and have permanent dipoles in the direction parallel to the long molecular axis. The MDC in this case can be written as a function of the order parameter, the molecular area, the distance betweeen the two electrodes, and the working area of the two electrodes. The details will be discussed in chapter 3.

1.4 Molecular Dynamics of Monolayer Films

1.4.1 Orientational Orders in Two-dimensional Systems

Despite the rapid experimental advancement, the interpretation of these experimental results at the beginning stage was surrounded by controversy, which emphasized the need for building a bridge between the physical phenomena and the architecture of monolayer films. [13] Even now the theories that exist have not yet been developed far enough to permit adequate calculations of magnitudes in individual cases, although sometimes these give some qualitative insight into physical mechanism. Three-dimensional (3D) theories, which have been known for quite a long time and widely used in solid physics and liquid crystals (LCs), [14] however, appeared to be paralyzed while dealing with problems in the field of monolayer films, typical two-dimensional systems. For Langmuir films, this is more striking, as the monolayer structure varies in the monolayer compression process, a case that is almost impossible in bulk states. Often, the structural change in bulk states is no more than a trivial deformation from the system's equilibrium position. The conversion from three dimensions to two dimensions usually leads to a decrease of symmetry. This gives a possibility that some implicit physical parameters in 3D systems might become explicit in 2D systems, which may manifest some characteristics of ultrathin films. As a monolayer film usually consists of dipolar molecules, discussing the molecular orientation makes more sense than other physical quantities. This molecular orientation motion often manifests itself through the macroscopic parameters such as the orientational order parameters. [6] The orientational order parameters in fact statistically characterize the

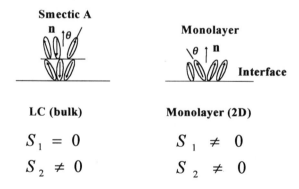

Figure 1.4 Structural difference between liquid crystals and monolayers.

orientational state of materials and the collective properties of molecules in mono-layers, such as those that occur in LCs, can be simply described by such kind of parameters. In liquid crystals, both the anti-ferroelectric and the para-ferroelectric states are possible due to the intermolecular interactions in the 3D structures. One generally uses the Maier-Saupe order, i.e., the thermal average of the second Legendre polynomial [15]

$$S_2 = \frac{3\langle \cos^2 \theta \rangle - 1}{2} \tag{1.1}$$

first introduced by Tsvetkov [11] into LCs, to represent the orientational state of molecules in liquid crystals. Here θ is the tilt angle of dipolar molecules from mono-layer normal, and $\langle \ \rangle$ stands for the thermal average. In the following discussions in this book we will use this notation with identical meaning to replace the notation of integral representing statistical average.

Similarly, when studying 2D systems of monolayer films on a material surface, it is desirable to find an order parameter to describe the molecular state. For monolayer films on a water surface, as the monolayer concerned is amphiphilic, the hydrophilic parts prefer to pin onto the surface due to the strong hydrophobic interaction, while the tails of these dipoles are free to rotate in the air. This is specified by the feature

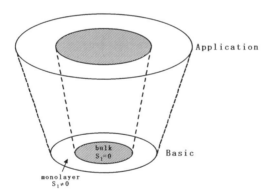

Figure 1.5 Organic monolayer dielectrics and electronics.

that the first orientational order parameter defined by thermal average of the first Legendre polynomial

$$S_1 = \langle P_1(\cos \theta) \rangle = \langle \cos \theta \rangle \tag{1.2}$$

is not zero and is usually used as the order parameter for monolayer films. For bulk dielectric materials such as liquid crystals, S_1 is zero even when the dielectric materials experience external stimulus. Thus we may expect that the presence of this unique order parameter S_1 in monolayers may lead to some new applications in the field of electronics (see Fig. 1.5). [16] In this book, the thermal average of Legendre polynomial $S_n = \langle P_n(\cos\theta) \rangle$ will be used as the notation for the orientational order parameters if without further explanation.

1.4.2 Interactions

The dipolar molecules in usual fluids, liquid crystals, and solids in bulk often reveal a self-assembly behavior after the various kinds of possible stereo configurations compete among themselves as a result of the intermolecular interactions. The elastic (curvature) free energy plays an important role in determining the relative stability of competing geometries of bulk materials. [14] In contrast, the one-layer structure of monolayer films makes this kind of interaction trivial compared with the electrostatic interactions.

There are two distinct interactions for molecules in monolayer films, the dipolar interaction and hard-core intraction. The dipolar interaction, which together with elastic deformation energy, governs the behaviors of liquid crystals, appears to be more important in monolayer films. Both parallel and antiparallel states of dipoles are possible in LCs, while dipoles in monolayer films prefer to line up in the same direction above material surface, due to the surface component of the interaction. Consequently, in a first order approximation, the magnitude of the dipole-dipole interaction in monolayer films is related to the first orientational order parameter rather than second orientational order parameter in LCs. Once this problem is clarified, it is evidently possible to establish connections between various physical properties and the first orientational order parameter with the help of some statistical approaches. Of course, this will inevitably involve the orientational distribution of the whole system, which depends on the interaction between molecules. It is assumed that this orientational distribution is ruled by Boltzmann statistics and will be discussed in more detail later in this book.

Another interaction that greatly influences the motion of the constituent molecules is the hard-core interaction. It is based on an assumption that dipolar molecules cannot interpenetrate into each other. When the molecular area is relatively small, the orientational motion of the molecular long axis is restricted within a conical angular range $\theta \in [0, \theta_A]$ due to the steric repulsive interaction among molecules. The more θ approaches θ_A, the stronger the steric repulsion force is.

1.4.3 State Equation and Piezoelectric Effect

The experimental surface pressure-area plateau observed for several decades in fact reveals the state equation of the molecular motions in monolayer films. The molecular motions within monolayer films include the 2D displacement and the orientational motion of the moelcular long axis. In a same way as that of an ideal gas, the surface pressure-area relation describes the motion state of the whole 2D monolayer system. Physically this relationship can be expressed by

$$\Pi = \gamma - \gamma_0 = \frac{\partial G}{\partial A}\big|_{T,P,n_i} - \gamma_0, \tag{1.3}$$

where G is the Gibss free energy, A is the molecular area, γ is the surface tension in the absence of a monolayer, γ_0 is the value with the monolayer present, and temperature T, pressure P and composition n_i are held constant. [1]

The piezoelectric effect in quartz was discovered over a century ago and the first transducer applications were suggested by Langevin in 1918. [1] Nowadays many applications of it have been achieved, such as nondestructive testing, pulse-echo transducers. Such kinds of phenomena usually occur around trivial deformation, that is, the crystal structure does not dramatically change. The most significant development of piezoelectric effect in ultrathin films came with the discovery of the piezoelectric phenomena of polyvinylidene fluoride (PVF$_2$) by Kawai. [17] Later, many investigations on the PVF$_2$ have been carried out. [18] The unique 2D structure of monolayer films gives a special importance to its specific piezoelectric effect from that in bulk state.

1.4.4 Dielectric Constant

The dielectric constant is an intricate coefficient reflecting the dielectric response abilities of dielectric materials to the influence of external electric fields. [19] Taylor and Bayes [20] gave an apparent dielectric constant definition accounting for the depolarization of the molecular dipoles for monolayer films on the basis of the Helmholtz model for unionized monolayers. [21] Furthermore, based on this definition, the influence of the orientation of the constituent molecules in monolayer films on the apparent dielectric constant can be carried out by considering a mean field produced by a 2D dipole array. [22] This can be traced back to the 1920s when Topping performed a local electric-field calculation of 2D array of rodlike uniaxial dipoles on a water surface. [23] The magnitude of this kind of electric field can be expressed as a function of some interaction constant. [24] Taking this into consideration, one should reflect the famous Madelung coefficient in the crystal lattice calculation, [25] which gives a simple indicator for the crystal structures. Later, Taylor and Bayes introduced an interaction constant g in the calculation of the dielectric constant of monolayers on a water surface. [20] However, what Taylor and Bayes defined is an apparent dielectric constant for monolayer films, not the real dielectric constant measured through experiment. As monolayer films consisting of constituent dipolar molecules are ferroelectric-like, one should in stead use differential definition. [19]

1.4.5 Phase Transition and Critical Phenomena

Phase transition and critical phenomena always attract physicists' attention. Over the last several decades, some experiments concerning phase transition phenomena in monolayer films have been achieved, though theoretically some still remained unclear for many years. [26] There are many kinds of phase transitions in monolayer films, for example, orientational phase transition,[19, 27] positional phase transition (pattern formation), [28] domain-forming transition, [13] phase transition during heating process, [29] chirality separation phase transition [30] and others. The orientational phase transition in Langmuir films as a result of monolayer compression will be discussed in Chapter 4. A complete theory explaining the domain formation has been given by McConnell *et al.* [13, 28] Chapter 8 introduces an interpretation for thermally stimulated phase transition. In Chapter 6, it is found that chiral monolayer films can be viewed as a Bragg-William binary cholesterics and the chirality separation phase transition is well explained by this point of view. While dealing with transition problems, methodologically Landau theory seems to have some advantages. [31] The mean field theory, as indeed it has been widely used in various fields of research study, [32] still reveals its advantage of reasonable simplicity in treating the phase transition [31] or the interactions between dipolar molecules in monolayer films, [7] as monolayer films have a C_∞ symmetry.

1.4.6 Dielectric Relaxation Phenomena

To understand the orientational state change of the molecules in monolayer films on a water surface during monolayer compression, one should address two questions: how does the monolayer compression affect the orientational state of the molecules and what is the most important determinant during this transient process. These are the dielectric relaxation phenomena. [33] Intuitively, these transient processes are predominantly hinged on a physical coefficient–relaxation time. [34] To describe the orientational motion of molecules in bulk materials, a comprehensive answer has been provided by the Debye rotational motion equation. [35] However, a new understanding of the dielectric relaxation time is needed, that is, it changes in the process of monolayer compression. Conventionally one treats the dielectric relaxation time (as in bulk states) as an invariable constant for a certain kind of material. But if we keep in mind that the relaxation time, an intricate parameter for a material, depends on the stereo molecular configuration in the material, there is nothing wrong in treating the relaxation time of monolayer films on a water surface as a continuous function of the molecular area, as the molecular structure varies with respect to the molecular area during the monolayer compression process. Once one gains information on the dielectric relaxation time, the dielectric dispersion properties of monolayer films can be estimated on the basis of Debye philosophy. [36]

1.4.7 Pattern Formation

As interesting topics, the chiral symmetry breaking (CSB), chiral discrimination (CD) and chiral phase separation (CPS) were extensively investigated.[37, 38] It is believed that in two-dimensional (2D) systems the question should be simplified [39]. Experimentally, a big amount of observations of pattern formation in 2D systems are believed to associate with CSB, such as in freely suspended films of smectic liquid crystals [40, 41], in Langmuir monolayers [41, 42, 43], and in Langmuir-Blodgett (LB) films [44]. In these experiments, Eckhardt *et al.*'s work [43] is quite notable. They studied the Langmuir monolayer of a kind of chiral tetracyclic alcohol, and observed three phases at different surface pressures. Using atomic force microscopy (AFM) under high pressure they imaged the formation of parallel stripes with alternate molecular packings, as well as large areas of uniform domains with mirror-symmetric positional orders. The existence of mirror symmetric positional orders, as well as the racemic composition of the monolayer, strongly implies the occurrence of CPS.

1.4.8 Optical Activity and Nonlinear Effect

There is no other subject as optical activity by which numerous studies can be carried out and to which many scientists pay so much attention. It is the asymmetric stereochemical architecture of molecules that is responsible for the phenomena of optical rotation and circular dichroism, as well as for a myriad of other physical and chemical properties. [45] Mathematically, optical activity itself is not a vectorial property but instead a property that has both sign and magnitude but not direction, i.e., a pseudoscalar. Methodologically, of course, a quantum mechanical treatment is needed. Much has been done in quantum chemistry concerning the photochemistry calculation. [46] All these calculations are performed for bulk state materials. Generally the electronic motion and the vibrational motion of the nucleus are considered using some kinds of approximation. The rotational motion part can be neglected, as the motion of molecules is usually restricted in the proximity of the equilibrium position. While in the monolayer films, the breaking of symmetry (3D→2D) allows the rotational motion of molecules within the restricted range $[0, \theta_A]$ and it appears to be more important than the vibrational one. This can be simply proved by the tran-cis isomerization phenomena in monolayer films observed through experiment. [9]

The *trans-cis* isomerization in monolayer films imposed by circularly polarized light is a much studied phenomenon in organic photochemistry. [47] Numerous experimental results concerning the *trans-cis* isomerization in monolayer films have been reported. [9, 48] At the same time, the theories that exist have not been developed far enough to permit adequate explanations concerning the individual experimental observations, such as photoizomerization reflected in mesurement of the Maxwell displacement current (MDC). [9]

How to describe a real molecular system and how to solve its Schrödinger equation have puzzled scientists for over sixty years. [49] The most inspiring advance in this field is the work of Born and Oppenheimer. They described the dynamics of the electrons through an expansion of the total wavefunction in adiabatic electronic wavefunctions, while the motion of the heavy nuclei was considered in a classical manner. This is called the Born-Oppenheimer (BO) adiabatic approximation. Later quite a number of heuristic models were proposed. Vibronic coupling, i.e. the interaction of nuclear vibration and electronic motion in molecules, [46] may cause a strong nonadiabatic coupling and the Born-Oppenheimer approximation is then no longer valid. A simple example is the Jahn-Teller effect. [50] They pointed out that the nuclei of molecules in degenerate electronic states would, in general, be unstable with respect to linear displacements along certain asymmetric normal coordinates. In this case, the Franck-Condon (FC) approximation is needed. The BO and FC approximations are confined to the harmonic approximation for vibrational motion and the one-particle approximation for electronic motion.

The many-body approach to the vibrational structure in molecular electronic spectra begins with the accurate calculation of FC factors for diatomic molecules, which were carried out using potential functions constructed by the Rydberg-Klein-Rees method. [51] However, this calculation is purely numerical. The theory proposed by Cerderbaum and Domcke [46] gives a simple formula to calculate FC factors. The experimental data concerning quantum dynamics of molecules require different interpretations of many-body systems. Usually a couple of motions in molecules are considered, such as vibrational, vibronic, torsional (for double-bond systems [52]) and rotational. A microscopic modeling of photoisomerization and internal-conversion dynamics has been introduced by Seidner and Domcke. [46] For a monolayer system, the rotational effect appears to be more important from the fact that the orientation of dipolar molecules changes as a result of photoirradiation. Chapter 7 focuses on the orientational effect of dipolar molecules.

The Hamiltonian of molecules is parity-invariant when it is confined to the framework of quantum electrodynamics. In the bulk state, it is difficult to observe the chirality as its effect will disappear as a result of three-dimensional averaging. However, some molecules (called *chiral molecules*), especially molecules in monolayer films, usually do not reveal parity invariance at all. This symmetry breaking, in general, will become explicit at the interface, for example in monolayer films, when exposed to photoirradiation due to the degeneracy of symmetry. The MDC generation, which does not occur in bulk state materials, is a good example of such symmetry breaking. The *trans-cis* photoisomerization phenomena in monolayer films, which arises from the orientational change of molecules, can be observed through MDC measurements. [8] The MDC detected in the experiment is in fact an average of the transition current over a certain time scale compared with the femto-second scale of electronic excitation.

Most phenomena concerning photoisomerization are real-time changes, or relaxational phenomena. [46, 53] The time-dependent Hamiltonian would be involved. Powerful computers have strongly reinforced some numerical approaches and enabled the straightforward numerical computation of the time-dependent Schrödinger equation. Often such problems should include nonlinear effects. Many researchers in the field of nonlinear optics relied on analyses in the frequency domain [54] so that a large amount of numerical calculation can be avoided. The photoizomerization MDC in the frequency domain will be discussed in a later chapter.

Whenever the interfacial properties are discussed, usually nonlinear effects have to be considered due the noncentrosymmetry at the interfaces. Many successful efforts have been made to describe the nonlinear optical activities, such as SHG, of noncentrosymmetric systems in the past several decades. [55] Physically SHG is a quantum mechanical phenomenon arising from the interaction between electrons in molecules and the applied light, [56] which is coherent during the interaction process. The SHG technique has been widely used to study the nonlinear optical response of some advanced materials such as fullerene. [57] On the other hand, when the dielectric response of monolayer films to the external electrostatic field, e.g. dc field or low-frequency field is discussed, it is also a behavior associated with molecules. Only static or low-frequency field can justify the neglect of delay of molecular reaction. The nonlinear effect in this case is called electro-optic effect, which is a statistical molecular phenomenon as well as an electronic and quantum mechanical polarization behavior.

SHG technique is based on the principle that a second-order process is forbidden in a medium with inversion symmetry under the electric-dipole approximation, but it is necessarily allowed at an interface between two bulk centrosymmetric media. Though the physical mechanism at an interface is actually very complicated and still partly remains unclear, the SH signals generated from the air-medium interface give much information on nonlinearity for the medium, such as nonlinear polarization. While on the other hand, MDC is a measurment technique associated with the zero order, i.e., the permanent dipole moment of constituent molecules. Therefore, it is quite natural that a combination of them will be very helpful for understanding the dielectrics of a material. A simultaneous measurement system of SHG and MDC will be detailedly discussed in Chapter 7.

1.4.9 Thermally Stimulated Phenomena

Thermally stimulated phenomena have been known for many years in various special forms such as thermally stimulated polarization (or depolarization) and thermally stimulated current (TSC). [58] The vitality of the investigation on thermal behaviors lies in the potential applications in the manufacturing process of thin films such as heterostructures. [59] Often a post-deposition thermal anneal may be needed to reduce deposition and thermal stresses. Many studies of TSC for multi- and monolayer films on a substrate have been reported. In a similar way as studies on Langmuir films by monolayer compression, such kind of investigations will inevitably involve dielectric relaxation [60]. A theory for obtaining the dipolar relxation time from TSC has been developed several decades ago and has been applied to several bulk materials, such as polyethylene, PVF_2, and others. [61] With a consideration of Keesom interaction for monolayer films in liquid-crystalline phase, a model is introduced to explain the thermally stimulated phase transition in monolayers, for example L-α-dimyristoyliphatidylcholine (DMPC) monolayer films in Chapter 8. Here again the orientational order parameter S_1 makes a significant contribution to the TSC, which cannot be seen in bulk materials.

1.4.10 Electronic Properties at MIM Interfaces

The interfacial electronic phenomena occurring at the metal/film and film/film interface or nanometric interfacial phenomena, to give a name that is currently more fashionable, has been a continous subject since the discovery of contact electrification phenomena. [62] The importance of these studies lies in the fact that the ultra thinness of films gives rise to the anisotropy of the materials of interest. The electronic properties, particularly in the direction perpendicular to the interface, is no longer an ensemble behavior and is strongly associated with its depth. Of interest are the electronic properties specific at interfaces, which usually fall into two main categories, the interfacial electrostatic phenomena and the dynamics behaviors such as electronic injection and tunnelling current. Such unique phenomena at interfaces, give great opportunities for scientists and engineers to find their potential applications in the field of nano electronic devices. As yet, no one has found an industrially valuable and technically revolutionary substitute for conventional microsized or submicrosized electronic devices based on silicon and lithium niobate. However, many successful attempts have been made at laboratory level. [63]

To clarify the interfacial electronic properties, it is crucially important to fabricate electronic elements consisting of pinhole-free ultra thin films without destroying their textures, which used to puzzle researchers in this field for quite a long time. Fortunately recent sophisticated techniques, such as LB deposition technique, have enabled the authors to achieve this goal. For example, the electrically insulating polyimide (PI) LB films with a monolayer thickness of 0.4 nm have been successfully fabricated by using a precursor method coupled with the conventional LB technique. [64] Here it is crucially important to keep in mind the origins of the interfacial electronic phenomena, where the displacement of electronic charges at the material/material interface and the orientational ordering of polar molecules are the main contributors and should be distinguished in the discussion on the interfacial phenomena.

Before the interests of nanometric phenomena were engaged, some techniques such as the heat-pulse-propagation technique, the pressure pulse technique, the electron beam method, and the electric stress-pulse technique had an advantage in that the space charge distribution can be measured without destroying the materials measured. [65] But as the resolution of these measurments is limited by the sound velocity of the bulk materials and measurably confined to micrometer scale, new techniques are desired. There are quite a number of measurement techniques available nowadays for studying the electronic properties at interfaces. Among them are X-ray Photoelectron Spectroscopy (XPS), [66] Ultraviolet Photoelectron Spectroscopy (UPS), [67] Photoemission Yield Spectroscopy (PEYS), [67] and Surface Potential Method (SPM) using Kelvin probe. [68] Using SPM, an investigation on electrical insulator PI LB films and semiconductor phthalocyanine LB films such as Cu-tetra-(tertbutyl)-phthalocyanine (CuttbPc) on various metal electrodes as a function of the number of deposited layers, revealed the presence of electron acceptor and donor states and excessive electronic charges transferred from metals to these LB films at the metal/LB film interfaces. [69]

It was also found that very high density of electronic states with an order of $10^{25} - 10^{26}$ m^{-3} and very high electric field with an order of $10^8 - 10^9$ V/m exist in the interfacial space charge layer within an order of several nanometers at the interface. Similar results were also obtianed for phthalocyanine ultrathin films prepared at very slow deposition rate by organic molecular beam evaporation technique. [70] A further analysis of their capacitance-voltage (C-V) and current-voltage (I-V) characteristics confirms the interfacial effect on the electrical transport properties of PI and phthalocyanine LB films. [71] The details will be provided in Chapter 9.

When the thickness of organic thin films as a kind of electrically insulating tunneling barriers is less than ten nanometers, it is possible to tunnel electrons across the films and a tunneling junction is formed. Such an electronic element requires the capability of preparing pinhole-free ultrathin films such as PI LB films. It is found that such films can be thermally and chemically stable up to a temperature of 400°C. Their electrical resistance is very high, usually greater than 10^{15} Ωcm, and their electrical breakdown strength is higher than 10^7 V/cm. These characteristics make the PI LB films work as good electrical insulating barriers in metal/insulator/metal (MIM) structures, as well as in tunneling junctions such as metal/barrier/super conductor and Josephson junctions. Other important application concerning use of ultrathin films as a tunneling barrier is the molecular rectifying junctions, electron resonance tunneling devices, and single electron tunneling devices. [72] For example, using a two-layer system consisting of octasubstituted palladiumphthalocyanine (PcPd) and perylene-tetra-carboperylene-tetra-carboxyldiimide derivative such as PTCDI-Spent and PTCDI-OET LB films, new electronic molecular diodes have been created on the basis of the asymmetric tunneling through molecular states, such as Highest Occupied Molecular Orbital (HOMO) and Lowest Unoccupied Molecular Orbital (LUMO) states of these molecules. [73] The tunneling phenomena of MIM structure are classified into elastic tunneling and inelastic tunneling effect, of which the electronic transport mechanism can be investigated via functionalized molecules in the artificially arranged multilayer films. Using PI LB films containing porphyrin (PORPI), elastic electron tunneling process and inelastic tunneling process of ultarathin films via porphyrin molecules have been found, [74] and the increase of electron tunneling current due to the excitation of electronic transition in molecular states in Q bands of porphyrins has been revealed in the inelastic tunneling spectra of junctions with a structure of Au/PI/PORPI/PI/Pb (or Pb-Bi, Au) at a temperature of 4.2 K. Similar experiments were carried out for Au/PI/rhodamine-dendorimer/PI/Au (or Al) junctions using rhodamine-dendorimer, showing a step structure in the I-V characteristic, which is similar to Coulomb staircase. Recently the possibility of single electron tunneling process via rhodamine molecule as a quantum dot has generated much interest.

Not surprisingly, the successful preparation of ultrathin films will find their various applications in the field of molecular electron devices, which is undoubtedly the most important and exciting field of electronic engineering in the 21st century. Of course the successful preparation of PI LB films is also inevitable in other fields such as liquid crystal display, where the ultrathin films are used as an alignment layer. [75]

References

[1] G. Roberts, *Langmuir-Blodgett Films*, Plenum, New York (1991); G. L. Gaines, Jr., *Insoluble Monolayers at Liquid-gas Interface*, Wiley-Interscience, New York (1965); M. C. Petty, *Langmuir-Blodgett Films*, Cambridge University Press, Cambridge (1996); A. Ulman, *Ultathin Organic Films*, Academic Press, New York (1991); K. Fukuda and M. Sugi, *Langmuir-Blodgett Films* Volumes 1-3, Elsevier, London (1989).

[2] M. Sugi, *Structure-dependent Carrier Transport in Langmuir Multilayer Assembly Films*, Researches of the Electrotechnical Laboratory, No. 794 (1978).

[3] M. N. Jones, *Micelles, Monolayers, and Biomembranes*, Wiley-Liss, New York (1995).

[4] V. K. Agarwal, *Electrical Behavior of Langmuir Films*, Electrocomponent Science and Technology, Gordon and Breach, New York (1975).

[5] G. Binnig and H. Rohrer, *Helv. Phys. Acta*, **55** (1982) 726.

[6] A. Sugimura, M. Iwamoto, and Z. C. Ou-Yang, *Phys. Rev.*, **E50** (1994) 614.

[7] C. X. Wu and M. Iwamoto, *Phys. Rev.*, **E57** (1998) 5740.

[8] M. Iwamoto and Y. Majima, *J. Chem. Phys.*, **94** (1991) 5135.

[9] M. Iwamoto, Y. Majima, H. Naruse, T. Noguchi, and H. Fuwa, *Nature*, **353** (1991) 645.

[10] S. Chandrasekhar, *Liquid Crystals*, Cambridge, London (1977).

[11] V. Tsvetkov, *Acta Physicochim. (USSR)*, **16** (1942) 132.

[12] J. Xue, C. S. Jung, and M. W. Kim, *Phys. Rev. Lett.*, **69** (1992) 474.

[13] G. M. Sessler, *Electrets*, Springer Verlag, New York (1987); D. J. Keller, H. M. McConnell, and V. T. Moy, *J. Phys. Chem.*, **90** (1986) 2311; A. Valance and C. Misbah, *Phys. Rev.*, **E55** (1997) 5564.

[14] P. G. de Gennes, *The Physics of Liquid Crystals*, Clarrendon, Oxford (1991).

[15] A. Saupe, *Z. Natureforsch.*, **19a** (1964) 161.

[16] M. Iwamoto, *IEICE Trans. Electron.*, in press (2000).

[17] H. Kawai, *Jpn. J. Appl. Phys.*, **8** (1969) 975.

[18] G. R. Davies, *Physics of Dielectric Solids*, Institute of Physics Conf. Series No. 58 (1980) p. 50; P. Pantelis, *Phys. Technol.*, **15** (1984) 239; G. M. Sessler, *J. Accoust. Soc. Am.*, **70** (1981) 1596.

[19] C. X. Wu, Z. C. Ou-Yang, and M. Iwamoto, *J. Chem. Phys.*, **109** (1998) 4552.

[20] D. M. Taylor and G. F. Bayes, *Phys. Rev.*, **E49** (1994) 1439.

[21] D. M. Taylor, O. N. Oliveira, Jr., and H. Morgan, *J. Colloid Interface Sci.*, **139** (1990) 508; J. R. Macdonald and C. D. Barlow, Jr., *J. Chem. Phys.*, **39** (1963) 412; R. J. Demchak and T. J. Fort, Jr., *J. Colloid Interface Sci.*, **46** (1974) 191.

[22] M. Iwamoto, Y. Mizutani, and A. Sugimura, *Phys. Rev.*, **B54** (1996) 8186; C. X. Wu and M. Iwamoto, *Phys. Rev.*, **B55** (1997) 10922.

[23] J. Topping, *Proc. R. Soc. London*, Ser. **A114** (1927) 67.

[24] R. E. Collin, *Field Theory of Guided Waves*, McGraw-Hill, New York (1960) Ch. 12.

[25] J. N. Israelachvili, *Intermolecular and Surface Forces*, Academic, London (1985).

[26] O. Albrecht, H. Gruler, and E. Sackmann, *J. Phys. (Paris)*, **39** (1978) 301; X. Qiu, J. Ruiz-Garcia, K. J. Stine, C. M. Knobler, and J. V. Selinger, *Phys. Rev. Lett.*, **67** (1991) 703.

[27] Z. Cai and S. A. Rice, *Faraday Discuss. Chem. Soc.*, **89** (1990) 211; Z. Cai and S. A. Rice, *J. Chem. Phys.*, **96** (1992) 6229.

[28] S. Perkovic and H. M. McConnell, *J. Phys. Chem.*, **101** (1997) 381.

[29] M. Iwamoto, C. X. Wu, and W. Y. Kim, *Phys. Rev.*, **B54** (1996) 8191.

[30] M. Iwamoto, C. X. Wu, and Z. C. Ou-Yang, *Chem. Phys. Lett.*, **285** (1998) 306.

[31] V. M. Kaganer and E. B. Loginov, *Phys. Rev. Lett.*, **71** (1993) 2599; V. M. Kaganer and E. B. Loginov, *Phys. Rev.*, **E51** (1995) 2237.

[32] F. Gießelmann and P. Zugenmaier, *Phys. Rev.*, **E55** (1997) 5613; J. Schacht, Gießelmann, P. Zugenmaier, and W. Kuczyński, *Phys. Rev.*, **E55** (1997) 5633.

[33] M. Iwamoto and C. X. Wu, *Phys. Rev.*, **E54** (1996) 6603.

[34] M. Iwamoto and C. X. Wu, *Phys. Rev.*, **E56** (1997) 3721.

[35] P. Debye, *Polar Molecules*, Dover, New York, (1929).

[36] H. Fröhlich, *Theory of Dielectrics*, Oxford University Press, New York (1958).

[37] L. Pasteur, *C. R. Acad. Sci. Paris*, **26** (1848) 535; A. Collet, M.-J. Brienne, and J. Jacques, *Chem. Rev.*, **80** (1980) 215; J. Jacques, A. Collet, and S. H. Wilen, *Enantiomers, Racemates and Resolutions*, Wiley, New York (1981); D. Andelman and P. G. de Gennes, *C. R. Acad. Sci.*, **307** (1988) 233; D. Andelman, *J. Am. Chem. Soc.*, **111** (1989) 6536; P. Nassoy *et al.*, *Phys. Rev. Lett.*, **75** (1995) 457.

[38] A. Buka and L. Kramer (editors), *Pattern Formation in Liquid Crystals*, Springer, New York (1996).

[39] M. V. Stewart and E. M. Arnert, in *Topics in Stereochemistry*, edited by N. L. Allinger, E. L. Eliel, and S. H. Wilen, Wiley, New York (1982).

[40] J. Maclennan and M. Seul, *Phys. Rev. Lett.*, **69** (1992) 2082; *Phys. Rev. Lett.*, **69** (1992) 3267; J. E. Maclennan, U. Sohling, N. A. Clark, and M. Seul, *Phys. Rev.*, **E49** (1994) 3207.

[41] D. K. Schwartz, *Nature (London)*, **362** (1993) 593.

[42] X. Qiu, J. Ruiz-Garcia, K. J. Stine, C. M. Knobler, and J. V. Selinger, *Phys. Rev. Lett.*, **67** (1991) 73; X. Qiu, J. Ruiz-Garcia, and C. M. Knobler, in *Interface Dynamics and Growth*, ed. K. S. Liang, M. P. Anderson, R. F. Bruinsma, and G. Scoles, Materials Research Society, Pittsburgh (1992), p. 263; F. Charra and J. Cousty, *Phys. Rev. Lett.*, **80** (1998) 1682.

[43] C. J. Eckhardt *et al.*, *Nature*, **362** (1993) 614.

[44] R. Viswanathan, J. A. Zasadzinski, and D. K. Schwartz, *Nature (London)*, **368** (1994) 440.

[45] J. Shao and P. Hänggi, *J. Chem. Phys.*, **107** (1997) 9935.

[46] L. Seidner and W. Domcke, *Chem. Phys.*, **186** (1994) 27; M. Seel and W. Domcke, *J. Chem. Phys.*, **95** (1991) 7806; L. S. Cederbaum and W. Domcke, *J. Chem. Phys.*, **64** (1976) 603; L. S. Cederbaum and W. Domcke, *J. Chem. Phys.*, **64** (1976) 612; G. Herzberg and H. C. Longuet Higgins, *Discussions Faraday Soc.*, **35** (1963) 77; R. L. Fulton and M. Gouterman, *J. Chem. Phys.*, **35** (1961) 1059.

[47] J. Michl and V. Bonačić-Koutecký, *Electronic aspects of organic photochemistry*, Wiley, New York (1990).

[48] W. Y. Kim and M. Iwamoto, *Mol. Cryst. Liq. Cryst.*, **280** (1996) 235.

[49] M. Born and R. Oppenheimer, *Ann. Physik. Leipzig*, **84** (1927) 457.

[50] H. A. Jahn and E. Teller, *Proc. Roy. Soc. (London)*, **A161** (1937) 22.

[51] W. Benesch, J. T. Vanderslice, S. G. Tilford, and P. G. Wilkinson, *Astrophys. J.*, **143** (1966) 236.

[52] G. J. M. Dormains, G. C. Groenenboom, and H. M. Buck, *J. Chem. Phys.*, **86** (1987) 4895.

[53] U. Müller and G. Stock, *J. Chem. Phys.*, **107** (1997) 6230.

[54] D. L. Mills, *Nonlinear Optics*, Springer, New York (1998).

[55] Y. R. Shen, *The Principles of Nonlinear Optics*, Wiley, New York (1984); N. Bloembergen and P. S. Pershan, *Phys. Rev.*, **128** (1962) 606.

[56] T. L. Mazely and W. M. Hetherington III, *J. Chem. Phys.*, **86** (1987) 3640; J. D. Byers, H. I. Yee, T. Petralli-Mallow, and J. M. Hicks, *Phys. Rev.* **B49** (1994) 14643; T. Petralli-Mallow, T. M. Wong, J. D. Byers, H. I. Yee, and J. M. Hicks, *J. Phys. Chem.*, **97** (1993) 1383.

[57] H. Huang, G. Gu, S. Yang, J. Fu, G. K. L. Wong, and Y. Du, *Chem. Phys. Lett.*, **272** (1997) 427.

[58] A. K. Jonscher, *J. Phys. D: Appl. Phys.*, **24** (1991) 1633.

[59] S. E. Rosenberg, P. Y. Wang, and I. N. Miaoulis, *Thin Solid Films*, **269** (1995) 64; S. Isomae, M. Nanba, Y. Tamaki, and M. Maki, *Appl. Phys. Lett.*, **30** (1977) 564.

[60] C. X. Wu, Y. Mizutani, and M. Iwamoto, *Jpn. J. Appl. Phys.*, **36** (1997) 222; M. Iwamoto, C. X, Wu, and Y. Mizutani, *J. Appl. Phys.*, **83** (1998) 4891.

[61] C. Bucci, R. Fieschi, and G. Guidi, *Phys. Rev.*, **148** (1966) 816; A. C. Lilly, Jr., R. M. Henderson, and P. S. Sharp, *J. Appl. Phys.*, **41** (1970) 2001; A. C. Lilly, Jr., L. L. Stewart, and R. M. Henderson, *J. Appl. Phys.*, **41** (1970) 2007; M. M. Perlman, *J. Appl. Phys.*, **42** (1971) 2645; T. Hino, *IEEE Trans. Electr. Insul.*, **21** (1986) 1007; T. Hino, *J. Appl. Phys.*, **48** (1976) 816; T. Hino, *J. Appl. Phys.*, **46** (1975) 1956; M. Iwamoto, K. Kato, and T. Hino, *IEEE Trans. Electr. Insul.*, **22** (1987)413, 425.

[62] J. Lowell and A. C. Rose-Innes, *Advances in Physics*, **29** (1980) 947, and references cited therein; L. H. Lee, *J. Electrostatics*, **32** (1994) 1, and references cited therein.

[63] T. Nakamura, in *Handbook of Organic Conductive Molecules and Polymers*, Vol.1 *Charge Transfer Salts, Fullerenes and Photoconductors*, ed. H. S. Nalwa, Wiley, New York (1997) Chap. 14, p. 727; S. M. Sze, *Physics of Semiconductor Devices*, Wiley, New York (1981).

[64] M. Iwamoto and M. Kakimoto, *Polyimides as Langmuir-Blodgett Films* in *Polyimides, Fundamentals and Applications*, ed. M. K. Ghosh and K. L. Mittal, Marcel Dekker, New York (1996); M. Suzuki, M. Kakimoto, T. Konishi,Y. Imai, M. Iwamoto, and T. Hino, *Chem. Phys. Lett.*, **395** (1986).

[65] G. M. Sessler, *IEEE Trans. Dielectrics and Electr. Insul.*, 4 (1997) 614; N. H. Ahmed and N. N. Srinvas, *IEEE Trans. Dielectrics and Electr. Insul.*, 4 (1997) 644.

[66] W. R. Salaneck, S. Stafström, and J. L. Bédas, *Conjugated Polymer Surfaces and Interfaces: Electronic and Chemical Structure of Interfaces for Polymer Light Emitting Devices*, Cambridge University Press, Cambridge (1996).

[67] F. Gutmann and L. E. Lyons, *Organic Semiconductors*, Wiley, New York (1967); K. Seki, in *Optical Techniques to Characterize Polymer Systems*, ed. H. Baessler, Elsevier, Amsterdam (1989).

[68] R. H. Tredgold and G. W. Smith, *IEEE Trans. Dielectrics and Electr. Insul.*, 4 (1997) 644; R. H. Tredgold and G. W. Smith, *Thin Solid Films*, **99** (1983) 215.

[69] E. Itoh and M. Iwamoto, *J. Appl. Phys.*, **81** (1997) 1790; E. Itoh, H. Kokubo, S. Shouriki, and M. Iwamoto, *J. Appl. Phys.*, **83** (1998) 372.

[70] Y. Majima, K. Yamagata, and M. Iwamoto, *J. Appl. Phys.*, **86** (1999) 3229.

[71] E. Itoh and M. Iwamoto, *J. Appl. Phys.*, **85** (1999) 7239; C. Q. Li, C. X. Wu, and M. Iwamoto, *Jpn. J. Appl. Phys.*, **39** (2000) 1840.

[72] W. Schutt, H. Koster, and G. Zuther, *Thin Solid Films*, **31** (1976) 275; C. M. Fischer, M. Burghard, S. Roth, and K. V. Klitzing, *Europhys. Lett.*, **28** (1994) 129.

[73] C. M. Fischer, M. Burghard, S. Roth, and K. V. Klitzing, *Appl. Phys. Lett.*, **26** (1995) 3331.

[74] M. Iwamoto, M. Wada, and T. Kubota, *Thin Solid Films*, **244** (1994) 472.

[75] W. Y. Kim, M. Iwamoto, and K. Ichimura, *Jpn. J. Appl. Phys.*, **35** (1996) 5395.

CHAPTER 2

POLARIZATION AND DIELECTRIC CONSTANT FOR 2D MEDIA

2.1 Polarization

Whenever we take the dielectric properties of organic mateials into consideration, it is no more than a reflection of the way the molecules of the medium are polarized by the local electric field. Such a local field is produced by the external electric field and the polarization of the molecules in the vicinity. The polarization per molecule of monolayer films can be classified into spontaneous polarization, first order polarization and nonlinear polarization, which is given by

$$\boldsymbol{p} = \boldsymbol{\mu} + \alpha \cdot \boldsymbol{E} + \chi : \boldsymbol{EE}. \tag{2.1}$$

Here $\boldsymbol{\mu}$ is the permanent dipole moment corresponding to the main contribution to the spontaneous polarization, α is the first-order polarizability and χ is the nonlinear polarizability. As we shall see later, Maxwell displacement current is attributed to the orientational change of dipolar molecules, dielectric constant mainly comes from the first-order polarization, and SHG and Pockels effect are strongly connected with χ. Considering the electronic polarization and orientational polarization, the total polarizability for isotropic bulk materials can be given by Debye-Langevin equation [1]

$$\alpha = \alpha^e + \frac{\mu^2}{3kT}, \tag{2.2}$$

where α^e is the linear electronic polarizability, k is the Boltzmann constant, and T is the temperature. The last term in Eq. (2.2) represents the contribution by the Boltzmann-averaged orientations of the dipolar molecules in isotropic bulk materials. However, for monolayers, this expression should be modified due to the symmetric breaking (see Fig. 1.4).

2.2 Spontaneous Polarization

Arising from the noncentrosymmetry at the interfaces (see Fig. 2.1), organic monolayers have spontaneous polarization $\boldsymbol{P}_0$ given by

$$\boldsymbol{P}_0 = N S_1 \mu \boldsymbol{n}, \tag{2.3}$$

where N is the number of molecules per molecular area, S_1 is the orientational order parameter defined as Eq. (1.2), μ is a permanent dipole moment of the molecule, and

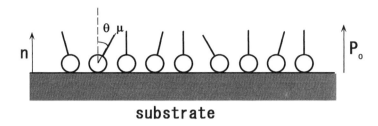

Figure 2.1 Model of polar molecules on the surface of material.

n is the normal direction of the monolayers. Such a polarization generates a Maxwell displacement current while the orientation of molecules is changed by external stimuli such as monolayer compression.

2.3 First Order Polarization and Dielectric Constant

The dielectric constant is an intricate coefficient reflecting the dielectric response abilities of dielectric materials to the external electric fields. [2] Its properties are strongly connected with monolayer structure, which determines the internal electric field produced by polar molecules. The internal field calculation can resort to the 2D mean field theory, the study of which can be traced back to 1920s when Topping performed a local electric-field calculation of 2D array of rodlike uniaxial dipoles on a water surface. [3] This kind of electric field attributed to the lattice structures can be expressed as a function of some interaction constant. [4] Considering this, one should reflect the famous Madelung coefficient in the crystal lattice calculation, [1] which gives a simple indicator for the crystal structures.

For monolayers, the calculation of the dielectric constant is also of considerable interest. Using a three-capacitor model, Taylor and Bayes [4] gave an apparent dielectric constant definition accounting for the depolarization of the dipolar molecules for monolayer films on the basis of the Helmholtz model with the Helmholtz equation (HE) for unionized (free from ions) monolayers. [5] It was defined as the ratio of the potential difference across the monolayers as a result of interactions to the potential difference across the monolayers without any interactions. [4] Furthermore, based on this definition and using a hard-core model, the influence of the orientation of the constituent molecules in monolayer films on the apparent dielectric constant can be calculcated. [6] Later, Taylor and Bayes introduced an interaction constant g in the calculation of the dielectric constant of monolayers on a water surface. [4] The definition as proposed by Taylor et al. played an important role in the study of, for example, the potential difference across monolayer films [5] and the Maxwell displacement current (MDC). [7, 8] A major and explicit advantage is that various physical quantities of monolayer system without external electric field can be simply expressed through such a constant. However, what Taylor and Bayes defined is an apparent dielectric constant for monolayer films, not the real dielectric constant measured through experiment. As monolayer films consisting of constituent dipolar molecules have spontaneous polarization, one should instead use differential definition, [2] i.e., a differentiation of the displacement field with respect to the external electric field at the point of zero external electric field. [9] This definition shall be adopted in this chapter. [10]

Let us consider a monolayer film system consisting of constituent molecules on a substrate or on a water surface. [10] The substrate or the water surface and the monolayer film itself can be considered as an infinite plane compared with the size of a single molecule of only several nanometers. We choose the coordinate system in such a way that the monolayer planes are parallel to the xy plane and the monolayer normal falls along the positive z axis, as shown in Fig. 2.2. Each molecule is assumed to have

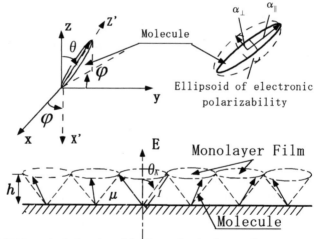

Figure 2.2 Molecular-orientational model. The molecules in monolayers are assumed to be uniaxial, depicted by a rotation ellipsoid of electronic polarizability.

a permanent dipole moment μ along its molecular long axis. The angle which the dipole moment at the origin makes with the layer normal is denoted by θ. The dipole at the origin discussed is assumed to be restricted within the angular range $[0, \theta_A]$ with $\sin^2 \theta_A = A/\pi l^2$ (A: molecular area, and l: the effective length of long molecular axis above the water surface), due to the hardcore repulsive interaction among molecules in monolayer films. Here it should be noted that as part of molecules may be more or less immersed in the water, we just consider the effective part above the water surface, possibly because the part immersed in the water is electrically screened compared with the part protruding in the air due to the high relative dielectric constant of water (78.6 at 25°C). The hardcore interaction produces no internal electric field in monolayer films, and can be regarded as an infinitely deep potential well between 0 and θ_A for the tilt angle θ. The molecular area decreases as the monolayer films are compressed. The molecule also orients with a counter-clockwise azimuthal angle φ from y axis. The coordinate system (x', y', z') is attached to the dipolar molecule at the origin with z' axis pointing toward the molecular long axis, and x' axis on the xy plane with a counter-clockwise angle φ away from x axis. Suppose the transformation from molecular frame (x', y', z') to laboratory frame (x, y, z) is represented by a tensor F, we have

$$\begin{pmatrix} x \\ y \\ z \end{pmatrix} = F \begin{pmatrix} x' \\ y' \\ z' \end{pmatrix},$$

where F is the transformation array representing the transformation from molecular frame (x', y', z') to laboratory frame (x, y, z) and given by

$$F = \begin{pmatrix} \cos\varphi & -\sin\varphi\cos\theta & -\sin\varphi\sin\theta \\ \sin\varphi & \cos\varphi\cos\theta & \cos\varphi\sin\theta \\ 0 & -\sin\theta & \cos\theta \end{pmatrix}. \tag{2.4}$$

2.3.1 Dielectric Anisotropy of Organic Monolayers

As the monolayer films we discuss are free from ions, it is quite adequate to consider solely the electronic polarization and the orientational polarization. Each kind of atoms or molecules is characterized by an electronic polarizability, which in most cases is not a scalar, but rather a tensor as a result of dielectric anisotropy. Often the electronic polarizability includes linear part and nonlinear parts. The nonlinear electronic polarizabilities originate from the quantum interaction between electrons and the external field, which can be expressed through a perturbation expansion. [11] Here we do not take the nonlinear parts higher than second order into account and assume that the polarization is isotropic in the plane perpendicular to the molecular long axis, i.e., the molecules are assumed to be dielectric uniaxial. In the molecular frame (x', y', z'), such an anisotropic electronic polarizability is distributed on a spin ellipsoid, and written by

$$\alpha_M^{(1)} = \begin{pmatrix} \alpha_\perp & 0 & 0 \\ 0 & \alpha_\perp & 0 \\ 0 & 0 & \alpha_\parallel \end{pmatrix}, \tag{2.5}$$

where $\alpha_\parallel$ and $\alpha_\perp$ are the electronic polarizability parallel and perpendicular to the molecular long axis respectively. In the laboratory frame (x, y, z), the description of the electronic polarizability can be obtained through a transformation

$$\alpha^{(1)} = F\alpha_M^{(1)}F^{-1}. \tag{2.6}$$

More specifically, its elements are given by

$$
\begin{aligned}
\alpha_{xx}^{(1)} &= \alpha_\perp(\cos^2\varphi + \sin^2\varphi\cos^2\theta) + \alpha_\parallel\sin^2\varphi\sin^2\theta \\
\alpha_{yy}^{(1)} &= \alpha_\perp(\sin^2\varphi + \cos^2\varphi\cos^2\theta) + \alpha_\parallel\cos^2\varphi\sin^2\theta \\
\alpha_{zz}^{(1)} &= \alpha_\perp\sin^2\theta + \alpha_\parallel\cos^2\theta \\
\alpha_{xy}^{(1)} &= \alpha_{yx}^{(1)} = -\Delta\alpha^{(1)}\sin\varphi\cos\varphi\sin^2\theta \\
\alpha_{xz}^{(1)} &= \alpha_{zx}^{(1)} = -\Delta\alpha^{(1)}\sin\varphi\sin\theta\cos\theta \\
\alpha_{yz}^{(1)} &= \alpha_{zy}^{(1)} = \Delta\alpha^{(1)}\cos\varphi\sin\theta\cos\theta,
\end{aligned}
\tag{2.7}
$$

where $\Delta\alpha^{(1)} = \alpha_\parallel - \alpha_\perp$, reflecting the linear dielectric anisotropy of electronic polarization.

On the other hand, the nonlinear electronic polarizability $\alpha_{M,i'j'k'}^{(2)}$ in the molecular frame (x', y', z') can be transformed to a macroscopic form $\langle\alpha_{ijk}^{(2)}\rangle$ in the laboratory frame (x, y, z) through the coordinate transformation tensor T [12]

$$\langle\alpha_{ijk}^{(2)}\rangle = \langle T_{ii'}T_{jj'}T_{kk'}\rangle\alpha_{M,i'j'k'}^{(2)}, \tag{2.8}$$

neglecting local-field effects, where $T = F$. For monolayer films with $C_{\infty v}$ symmetry, as the orientational distribution is isotropic in the plane of the monolayer surface, Eq. (2.8) yields three independent elements of $\alpha^{(2)}$ in terms of S_1 and S_3: [12]

$$
\begin{aligned}
\langle\alpha_{zzz}^{(2)}\rangle &= \frac{1}{5}\alpha_{M,z'z'z'}^{(2)}(3S_1 + 2S_3) + \frac{1}{5}(\alpha_{M,z'x'x'}^{(2)} + 2\alpha_{M,x'z'x'}^{(2)})(S_1 - S_3) \\
\langle\alpha_{zxx}^{(2)}\rangle &= \alpha_{zyy}^{(2)} = \frac{1}{20}(2\alpha_{M,z'z'z'}^{(2)} - \alpha_{M,z'x'x'}^{(2)} - 2\alpha_{M,x'z'x'}^{(2)})(S_1 - S_3) + \alpha_{M,z'x'x'}^{(2)}S_1 \\
\langle\alpha_{xzx}^{(2)}\rangle &= \alpha_{yzy}^{(2)} = \frac{1}{20}(2\alpha_{M,z'z'z'}^{(2)} - \alpha_{M,z'x'x'}^{(2)} - 2\alpha_{M,x'z'x'}^{(2)})(S_1 - S_3) + \alpha_{M,x'z'x'}^{(2)}S_1,
\end{aligned}
\tag{2.9}
$$

assuming that in the molecular frame (x', y', z'), only three independent, nonvanishing elements, $\alpha_{M,z'z'z'}^{(2)}$, $\alpha_{M,z'x'x'}^{(2)}$, and $\alpha_{M,x'z'x'}^{(2)} = \alpha_{M,x'x'z'}^{(2)}$, of second-order electronic polarizability tensor exist. If $\alpha_{M,x'z'x'}^{(2)} \neq \alpha_{M,x'x'z'}^{(2)}$, one gets four independent nonvanishing elements in the laboratory frame for systems with $C_{\infty v}$ symmetry. [13] Here S_1 and S_3 are the first and the third orientational order parameters in the absence of the external field. We use $\langle\cos^2\varphi\rangle = 1/2$ and rewrite the results by Heinz [12] in terms of the orientational order parameters S_1 and S_3 in a similar way as that by Ou-Yang and Xie. [14]

Now we consider a system of molecules in a monolayer film being subjected to an external electrostatic or low-frequency field $\boldsymbol{E} = (E_x, E_y, E_z)$. Such an external field will induce a polarization $\boldsymbol{p} = \boldsymbol{\mu} + \alpha^{(1)} \cdot \boldsymbol{E} + \alpha^{(2)} : \boldsymbol{EE}$, where $\alpha^{(1)}$ is the linear polarizability and $\alpha^{(2)}$ is the second-order susceptibility. The dipolar molecule at the origin experiences a total electric field of $(E_x, E_y, E_z + E_z^{in})$, where E_z^{in} is the internal electric field produced by the molecules in the vicinity. As the permanent dipole moment μ coincides with the molecular long axis, it is expressed as $\boldsymbol{\mu} = (-\mu \sin\theta \sin\varphi, \mu \sin\theta \cos\varphi, \mu \cos\theta)$ in the laboratory frame. The interaction energy between the dipole and the external field is supposed to be

$$
\begin{aligned}
W & = -\int (\boldsymbol{E} + E_z^{in}\boldsymbol{e}_z) \cdot d\boldsymbol{m} \approx -\boldsymbol{\mu} \cdot (\boldsymbol{E} + E_z^{in}\boldsymbol{e}_z) \\
& = \mu E_x \sin\theta \sin\varphi - \mu E_y \sin\theta \cos\varphi - \mu(E_z + E_z^{in})\cos\theta,
\end{aligned}
\tag{2.10}
$$

where $\boldsymbol{m}$ is the polarization, and $\boldsymbol{e}_z$ is the z-direction unit vector in the laboratory frame. It is reasonable to neglect the interaction energy part between the external field and the induced dipoles due to the electronic polarization $\alpha^{(1)}$ and $\alpha^{(2)}$ in Eq. (2.10), because they are proportional to E_i^2 or E_i^3 $(i = x, y, z)$ and as a whole only effect the third-order or four-order nonlinearity after an expansion of $e^{-W/kT}$ and an integration over all possible orientational position in the later calculations [Eq. (2.11)]. In order to calculate the polarization of the monolayer film, an average of the electronic polarizability Eq. (2.6) over all possible orientational position (φ, θ) of the molecule at the origin is necessary. The result can be written as

$$
\begin{aligned}
\langle \alpha^{(1)} \rangle & = \frac{1}{Z} \int_0^{2\pi} \int_0^{\theta_A} \alpha^{(1)} e^{-W/kT} d\varphi \sin\theta d\theta \\
& = \overline{\alpha^{(1)}} I - \frac{1}{3}\Delta\alpha^{(1)}\Gamma S_2 + \alpha_{NL}^{(1)},
\end{aligned}
\tag{2.11}
$$

where $\overline{\alpha^{(1)}} = (\alpha_\parallel + 2\alpha_\perp)/3$ is the linear electronic polarizability average representing the isotropic contribution, I is a 3×3 unit tensor, $\alpha_{NL}^{(1)}$ is the nonlinear part of the polarizability produced by the local field, and Z is the partition function given by

$$
Z = \int_0^{2\pi} \int_0^{\theta_A} e^{-W/kT} d\varphi \sin\theta d\theta.
\tag{2.12}
$$

We discard all the higher-order expansion terms than $E_i (i = x, y, z)$ during the calculation of Eq. (2.11), as they only cause the third and the higher orders of polarization $\boldsymbol{p}$. In order to discuss the dielectric anisotropy, we introduce a traceless anisotropic order parameter tensor S_{ani}

$$
S_{ani} = \Gamma S_2 = \begin{pmatrix} 1 & 0 & 0 \\ 0 & 1 & 0 \\ 0 & 0 & -2. \end{pmatrix} S_2
\tag{2.13}
$$

to represent the influence of orientational order on the electronic anisotropy of molecules. Here S_2 is the second orientational order parameter in the absence of external electric field defined as Eq. (1.1). If we keep in mind that the monolayer film discussed has a $C_{\infty v}$ symmetry, there is nothing strange that the electronic polarization should remain unchanged under a permutation operation $x \leftrightarrow y$ and an inversion operation $x \rightarrow -x$ or $y \rightarrow -y$. Such structure symmetries of monolayer films immediately make S_{ani} diagonal and $S_{ani11} = S_{ani22}$, as we can see in Eq. (2.13).

The dielectric anisotropy of the monolayer films arises from the anisotropy of a single molecule's electronic polarizability. Such a difference between the polarizations parallel and perpendicular to the molecular long axis, after a Boltzmann average over all the molecules on an indefinite plane, leads to an additional collective term, i.e. the anisotropic part, to the dielectric constant. It is interesting from Eq. (2.13) that in bulk states, the anisotropy disappears as a result of an average over all possible orientational positions. While a field in z direction applied to a polar system will bring about a positive anisotropic effect of $2\Delta\alpha^{(1)} S_2/3$ on the electronic polarization, and a field in either x or y direction will cause an opposite effect but with half magnitude. It is important to note that the calculation of the anisotropic effect in this study is based on an assumption that the molecules are uniaxial, i.e., the polarizability distribution itself forms a rotational ellipsoid. In real cases, any deviation from the uniaxial assumption will lead to a varied, but similar result. Thus it is not included here.

Another characteristic of the orientational order S_{ani} responsible for the dielectric anisotropy is that S_{ani} does not depend on the first order S_1. Such a feature is attributed to the symmetry of the electronic polarization. The anisotropy of the electronic polarization discussed is assumed to come from the ellipsoidal orbit of electrons around their nuclear in the absence of the external electric field. Such a deviation from the isotropical spherical orbit, does not break the centrosymmetry of the monolayer film as a whole even after an orientational average performance. Thus, the anisotropic electronic polarization is only associated with the centrosymmetric second orientational order parameter S_2 rather than S_1 or S_3, which possesses noncentrosymmetry.

2.3.2 Dielectric Contribution by Molecular Orientation

When an external electrostatic or a low-frequency field $\boldsymbol{E} = (E_x, E_y, E_z)$ is applied to the monolayer film, the dipolar molecules prefer to reorient along the direction of the applied field due to the interaction given by Eq. (2.10). In order to discuss the nonlinear orientational influence of permanent dipole moments on the dielectric constant and for simplicity, it is assumed, quite reasonably, that the external field E_i is larger than the internal field E_z^{in} so that the influence of the molecule-molecule interaction on the polarization $\boldsymbol{m}$ can be neglected. Suppose the orientational motion of constituent molecules in the monolayer film obeys the Boltzmann distribution rule, the average of the dipole moment components is given by

$$\langle \mu_i \rangle = \frac{1}{Z} \int_0^{2\pi} \int_0^{\theta_A} \mu_i e^{-W/kT} d\varphi \sin\theta d\theta$$

$$= \frac{1}{Z} \int_0^{2\pi} \int_0^{\theta_A} \mu_i \left(1 - \frac{W_E}{kT} + \frac{W_E^2}{2k^2T^2} \right) e^{\mu E_z^{in} \cos\theta / kT} d\varphi \sin\theta d\theta, \quad (2.14)$$

where μ_i $(i = x, y, z)$ is the i-direction component of the permanent dipole μ and Z is the partition function. After a lengthy calculation, the dipole moment average can be written in a compact form:

$$\langle \boldsymbol{\mu} \rangle = \mu S_1 \boldsymbol{e}_z + \frac{\mu^2}{3kT} S_{ori} \cdot \boldsymbol{E} + \boldsymbol{\mu}_{NL}, \quad (2.15)$$

where $\boldsymbol{e}_z$ is the unit vector along z direction, and $\boldsymbol{\mu}_{NL}$ is the nonlinear orientational effect which will be discussed in later sections. Here we define a tensor

$$S_{ori} = \begin{pmatrix} 1 - S_2 & 0 & 0 \\ 0 & 1 - S_2 & 0 \\ 0 & 0 & 1 - 3S_1^2 + 2S_2 \end{pmatrix}, \quad (2.16)$$

to describe the orientational polarization. S_{ori} describes the influence of the orientational motion of polar molecules in monolayer films, which brings an additional term to the dielectric polarizability, as will be shown later. It is clear that S_{ori} depends on the monolayer structure characterized by the two orientational parameters S_1 and S_2. It is also found from Eq. (2.16) that the orientational polarization creates apparent dielectric polarizabilities

$$\alpha_{zz} = \alpha^e + \frac{\mu^2}{3kT}(1 - 3S_1^2 + 2S_2)$$

$$\alpha_{xx} = \alpha_{yy} = \alpha^e + \frac{\mu^2}{3kT}(1 - S_2) \quad (2.17)$$

in the normal direction and planar direction respectively. From Eq. (2.17), it is clear that in the case of bulk states, i.e. $S_1 = S_2 = 0$, $S_{ori} = I$, the orientational contribution becomes $\mu^2/3kT$. This is nothing but the Debye-Langevin equation Eq. (2.2).

The polar molecules in a weak external electric field tend to align approximately along the electric field, creating an induced dipole moment after a Boltzmann average over all polar molecules. The induced dipole moment depends on the electric field applied to the molecules. This is where the orientational polarization comes from. If we take the $C_{\infty v}$ symmetry of the monolayer film into account, it is not strange that the orientational polarization tensor as defined by Eq. (2.16) is diagonal. Moreover, if $S_1 = 0$, Eq. (2.16) is nothing but the result for liquid crystals.

2.3.3 Dielectric Constant

When an electrostatic field $\boldsymbol{E}$ is applied to a monolayer film, the monolayer film is polarized and a displacement field $\boldsymbol{D}$ is induced. It is composed of the applied field and the polarization

$$D = \epsilon_0 E + \frac{1}{Ah\epsilon_0}[\langle \mu \rangle + \langle \alpha^{(1)} \rangle \cdot E + \langle \alpha^{(2)} \rangle : EE]. \tag{2.18}$$

From the above equation, we immediately get the dielectric constant tensor

$$
\begin{aligned}
\epsilon_r &= \frac{\partial D_i}{\epsilon_0 \partial E_j} \\
&= I + \epsilon^e + \epsilon^{ori} + \epsilon_{NL} \\
&= I + \frac{1}{Ah\epsilon_0}[(\overline{\alpha^{(1)}}I + \frac{\mu^2}{3kT}S_{ori}) - \frac{1}{3}\Delta\alpha^{(1)}\Gamma S_2] + \epsilon^p_{ijk}E_k, \tag{2.19}
\end{aligned}
$$

where I is the unit tensor, $\epsilon^e = \frac{1}{Ah\epsilon_0}\overline{\alpha^{(1)}}I$ is the relative dielectric constant due to the electronic polarizability, $\epsilon^{ori} = \frac{1}{Ah\epsilon_0}\frac{\mu^2}{3kT}S_{ori}$ comes from the orientational motion of molecules, and $\epsilon^p_{ijk}E_k$ is the nonlinear eletrooptical Pockels effect and will be discussed later.

It is important to discuss the relationship between the internal electric field E_z^{in} and the first orientational order parameter S_1. Such an internal electric field is produced by the dipole moments in the vicinity and can be calculated by assuming an infinite array of molecules with an effective dipole moment of μS_1 on a 2D plane. Here we express it in a form of [2]

$$E_z^{in} = -g_z \mu a^{-3/2} S_1, \tag{2.20}$$

where g_z is a nonunit factor reflecting the positional order, i.e. the configuration of molecules in monolayer films and defined as

$$g_z = \sum_i \frac{a^{3/2}}{4\pi\epsilon_0 r_i^3}, \tag{2.21}$$

where a is the relative molecular area ($= A/A_0$) and r_i is the distance of molecules from the origin. The summation is carried out over all molecules except the one at the origin. For hexagonal packing, $g_z = 11.0342$. [4] Considering that S_1 itself can only be calculated when E_z^{in} is known, we have a self-consistent equation for calculating S_1

$$S_1 = \frac{1}{Z_\theta}\int_0^{\theta_A}\cos\theta e^{-g_z\mu a^{-3}S_1\cos\theta/kT}\sin\theta d\theta, \tag{2.22}$$

where Z_θ is the partition function with respect to θ. A similar way of calculation can also give S_2 and S_3. From Eqs. (2.13), (2.16) and the above equation, we know that the internal electric field indirectly influences the dielectric constant through the three orientational order parameters S_1, S_2 and S_3.

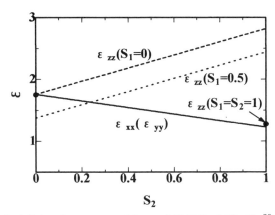

Figure 2.3 Typical dielectric constant with $\mu = 1$ D (1 D=3.33×10^{-30} C·m), $T = 300$ K, $A = 20$ Å^2, $h = 10$ Å, $\alpha_\parallel = 4.8$ Å^3, $\alpha_\perp = 3.6$ Å^3 for Eq. (2.19). Here we ignore the nonlinear contribution in the calculation.

Figure 2.3 shows a typical example of the dependence of the dielectric constant for monolayer films on the orientational order parameters S_1 and S_2, neglecting the nonlinear effect. We choose $\mu = 1$ D, $T = 300$ K, $A = 20$ Å^2, $h = 10$ Å, $\alpha_\parallel = 4.8$ Å^3, $\alpha_\perp = 3.6$ Å^3. It is found that ϵ_{zz} increases while $\epsilon_{xx}(\epsilon_{yy})$ decreases with the increase of S_2, because the alignment order along the normal (z) direction decreases the contribution of the in-plane (x or y-direction) polarization. Figure 2.3 also reveals ϵ_{zz} increases while S_1 decreases. This indicates that S_1 is a destructive order and S_2 is a constructive order to the dielectric polarization along the normal direction. This is pretty important because a dependence of dielectric constant on material structures is noted. And it is interesting to note that two extreme cases $\epsilon_{zz} = \epsilon_{xx} = \epsilon_{yy} = 1.76$ ($S_1 = S_2 = 0$) and $\epsilon_{zz} = 1.30$ and $\epsilon_{xx} = \epsilon_{yy} = 1.23$ ($S_1 = S_2 = 1$) corresponds to isotropic bulk materials and low temperature or sufficiently strong field respectively.

2.4 Nonlinear Polarization

Whenever the interfacial properties are discussed, usually nonlinear effects have to be considered due to the noncentrosymmetry at the interfaces. Many successful efforts have been made to describe the nonlinear optical activities, such as SHG, of noncentrosymmetric systems in the past several decades. [11, 6, 16] The SHG technique has been widely used to study the nonlinear optical response of some advanced materials such as fullerenes. [17] On the other hand, when we discuss the dielectric response of monolayer films to the external electrostatic or low-frequency field, it is also a behavior associated with molecules. Only static or low-frequency field can make us neglect the delay of molecular reaction. There have been literatures on the macroscopic properties in orientationally ordered materials. [12, 18, 19] However, except for the second-order nonlinear properties such as susceptibility, they gave no information on the coupling between first orders, such as dielectric anisotropy and orientational motions. The hyperpolarization arising from the nonlinear electronic polarizability $\langle \alpha^{(2)} \rangle$, nonlinear additional electronic polarizability $\alpha_{NL}^{(1)}$ produced by the local-field molecular interaction, and the nonlinear orientation-induced dipole moment μ_{NL} as a result of the interaction between the dipole moment and the external field, is given by

$$\boldsymbol{P}^{(2)} = \chi^p \cdot \boldsymbol{EE} = \frac{1}{Ah\epsilon_0}(\langle \alpha^{(2)} \rangle : \boldsymbol{EE} + \alpha_{NL}^{(1)} \cdot \boldsymbol{E} + \boldsymbol{\mu}_{NL}), \qquad (2.23)$$

where A is the area per molecule, h is the thickness of the monolayer, $\langle \alpha^{(2)} \rangle$ is the second-order electronic polarizability average neglecting the local-field effects expressed by Eq. (2.9). $\alpha_{NL}^{(1)}$ in Eq. (2.11) and Eq. (2.23) is written as

$$\alpha_{NL}^{(1)} = \begin{pmatrix} \gamma_2 E_z & 0 & \gamma_1 E_x \\ 0 & \gamma_2 E_z & \gamma_1 E_y \\ \gamma_1 E_x & \gamma_1 E_y & -2\gamma_2 E_z \end{pmatrix}, \qquad (2.24)$$

where

$$\begin{aligned} \gamma_1 &= \frac{\Delta\alpha^{(1)}\mu}{5kT}(S_1 - S_3) \\ \gamma_2 &= \frac{\Delta\alpha^{(1)}\mu}{15kT}(5S_1 S_2 - 2S_1 - 3S_3). \end{aligned} \qquad (2.25)$$

At the same time, the nonlinear orientation-induced dipole moment can be given by

$$\boldsymbol{\mu}_{NL} = \begin{pmatrix} \kappa_1 E_x E_z \\ \kappa_1 E_y E_z \\ \kappa_2(E_x^2 + E_y^2) + \kappa_3 E_z^2 \end{pmatrix}, \qquad (2.26)$$

where

$$\begin{aligned} \kappa_1 &= \frac{\mu^3}{15k^2T^2}(5S_1 S_2 - 2S_1 - 3S_3) + \frac{2\Delta\alpha^{(1)}\mu}{5kT}(S_1 - S_3) \\ \kappa_2 &= \left(\frac{\mu^3}{30k^2T^2} + \frac{\Delta\alpha^{(1)}\mu}{15kT}\right)(5S_1 S_2 - 2S_1 - 3S_3) \\ \kappa_3 &= \frac{\mu^3}{5k^2T^2}(5S_1^3 - 5S_1 S_2 - S_1 + S_3) - \frac{2\Delta\alpha^{(1)}\mu}{15kT}(5S_1 S_2 - 2S_1 - 3S_3). \end{aligned} \qquad (2.27)$$

For monolayer films, the $C_{\infty v}$ symmetry makes most of the elements of the $3 \times 3 \times 3$ nonlinear polarizability χ_{ijk}^p zero. Only three elements of χ_{ijk}^p are found to be independent as a result of the centrosymmetry in the xy plane. Substituting Eq. (2.24) and Eq. (2.26) into Eq. (2.23), we get

$$\chi^p_{ijk} = \frac{1}{Ah\epsilon_0}\langle\alpha^{(2)}_{ijk}\rangle + \delta\chi_{ijk} \tag{2.28}$$

with the additional orientation-induced nonlinear polarizability

$$\delta\chi_{ijk} = \frac{1}{Ah\epsilon_0}\begin{pmatrix} 0 & 0 & 0 & 0 & \beta_1/2 & 0 \\ 0 & 0 & 0 & \beta_1/2 & 0 & 0 \\ \beta_2 & \beta_2 & \beta_3 & 0 & 0 & 0 \end{pmatrix}, \tag{2.29}$$

where $\beta_1 = \kappa_1 + \gamma_1 + \gamma_2$, $\beta_2 = \kappa_2 + \gamma_1$, and $\beta_3 = \kappa_3 - 2\gamma_2$ with γ_1, γ_2, κ_1, κ_2 and κ_3 expressed as Eq. (2.25) and Eq. (2.27). From Eqs. (2.25), (2.27) and (2.29), it is shown that the additional nonlinear polarization is induced by the nonlinear orientational polarization of constituent dipolar molecules and the coupling between the anisotropy of electronic polarization and the orientational polarization. This is the nonlinearity due to local-field effects. Such an effect strongly depends on the permanent dipole moment μ. When $\mu = 0$, $\delta\chi_{ijk} = 0$.

From the nonlinear polarization Eq. (2.23) and Eq. (2.29), the nonlinear eletrooptical Pockels effect, which is given by

$$\epsilon^p_{ijk}E_k = \frac{2}{Ah\epsilon_0}\begin{pmatrix} (\langle\alpha^{(2)}_{xzx}\rangle + \beta_1)E_z & 0 & (\langle\alpha^{(2)}_{xzx}\rangle + \beta_1)E_x \\ 0 & (\langle\alpha^{(2)}_{xzx}\rangle + \beta_1)E_z & (\langle\alpha^{(2)}_{xzx}\rangle + \beta_1)E_y \\ (\langle\alpha^{(2)}_{zxx}\rangle + \beta_2)E_x & (\langle\alpha^{(2)}_{zxx}\rangle + \beta_2)E_y & (\langle\alpha^{(2)}_{zzz}\rangle + \beta_3)E_z \end{pmatrix}, \tag{2.30}$$

can be discussed. If we notice that the monolayer film system is noncentrosymmetric, it is natural that the nonlinearity is included into the dielectric constant calculation. Such structural characteristic, on the other hand, can be described by the orientational order parameters. Thus the Pockels effect should be able to be expressed in orientational orders. Combining Eqs. (2.25), (2.27), and Eq. (2.29), together with Eq. (2.9), leads to the three independent, nonvanishing elements of nonlinear polarizability $\chi^p_{ijk} = \chi_{ijk} + \delta\chi_{ijk}$:

$$\chi^p_{xzx} = \frac{1}{20Ah\epsilon_0}(2\alpha^{(2)}_{M,z'z'z'} - \alpha^{(2)}_{M,z'x'x'} - 2\alpha^{(2)}_{M,x'z'x'})(S_1 - S_3) + \frac{1}{Ah\epsilon_0}\alpha^{(2)}_{M,x'z'x'}S_1$$

$$+\frac{\Delta\alpha^{(1)}\mu}{30Ah\epsilon_0 kT}(5S_1S_2 + 7S_1 - 12S_3) + \frac{\mu^3}{30Ah\epsilon_0 k^2T^2}(5S_1S_2 - 2S_1 - 3S_3)$$

$$\chi^p_{zxx} = \frac{1}{20Ah\epsilon_0}(2\alpha^{(2)}_{M,z'z'z'} - \alpha^{(2)}_{M,z'x'x'} - 2\alpha^{(2)}_{M,x'z'x'})(S_1 - S_3) + \frac{1}{Ah\epsilon_0}\alpha^{(2)}_{M,z'x'x'}S_1$$

$$+\frac{\Delta\alpha^{(1)}\mu}{15Ah\epsilon_0 kT}(5S_1S_2 + S_1 - 6S_3) + \frac{\mu^3}{30Ah\epsilon_0 k^2T^2}(5S_1S_2 - 2S_1 - 3S_3)$$

$$\chi^p_{zzz} = \frac{1}{5Ah\epsilon_0}\alpha^{(2)}_{M,z'z'z'}(3S_1 + 2S_3) + \frac{1}{5Ah\epsilon_0}(\alpha^{(2)}_{M,z'x'x'} + 2\alpha^{(2)}_{M,x'z'x'})(S_1 - S_3)$$

$$-\frac{4\Delta\alpha^{(1)}\mu}{15Ah\epsilon_0 kT}(5S_1S_2 - 2S_1 - 3S_3)$$

$$+\frac{\mu^3}{5Ah\epsilon_0 k^2 T^2}(5S_1^3 - 5S_1S_2 - S_1 + S_3), \qquad (2.31)$$

where $\chi_{xzx}^p = \chi_{xxz}^p = \chi_{yzy}^p = \chi_{yyz}^p$, and $\chi_{zxx}^p = \chi_{zyy}^p$. From Eq. (2.30), the electro-optic Pockels effect ϵ_{ijk}^p are given by $\epsilon_{ijk}^p = 2\chi_{ijk}^p$. If $D_i = \epsilon_{ij}E_j$ is used rather than the differentiation Eq. (2.19) as the definition of the dielectric constant, then $\epsilon_{ijk}^p = \chi_{ijk}^p$. From Eq. (2.30), it is indicated that the nonlinear Pockels effect is attributed to the nonlinear electronic polarization $\alpha_{M,i'j'k'}^{(2)}(i = x, y, z)$ of a single molecule, the structural noncentrosymmetry, which can be quantized as orientational order parameters S_1 and S_3, the nonlinear orientational polarization ($\propto \mu^3$), and the coupling between anisotropy and the orientation ($\propto \Delta\alpha^{(1)}\mu$). All of these contributions also come from their noncentrosymmetric characteristic: $\alpha_{M,i'j'k'}^{(2)}(i, j, k = x, y, z)$ can be non-centrosymmetric; S_1 and S_3 in Eq. (2.30) are noncentrosymmetric; and the permanent dipole moment is also noncentrosymmetric. The dielectric anisotropy itself in this paper is centrosymmetric, but a coupling between the anisotropy and the permanent dipole moment shows noncentrosymmetry. The additional parts in Eq. (2.31) are respectively proportional to μ and μ^3, the signs of which change under an operation of spatial inversion. From Eq. (2.31) and Eq. (2.30), it is also found that in isotropic systems, $S_1 = S_2 = S_3 = 0$, which leads to $\chi_{xzx}^p = \chi_{yzy}^p = \chi_{zxx}^p = \chi_{zyy}^p = \chi_{zzz}^p = 0$, and then $\epsilon_{ijk}^p = 0$. This reveals that there is no Pockels effect in the isotropic systems with centrosymmetry. A combination of the Maxwell displacement current measurement, [20] the SHG detection and the above equations will help understand the physical quantities such as $\alpha_{M,i'j'k'}(i, j, k = x, y, z)$, which include chirality information, at a molecular level for monolayer films.

2.5 Summary

The spontaneous, linear and nonlinear polarizations of a monolayer film determine its dielectric behaviors. The spontaneous polarization induces a Maxwell displacement current in the normal direction by external stimulation such as monolayer compression. The dielectric constant mainly comes from linear contribution. A numerical calculation shows that the first orientational order parameter S_1 is a destructive order while the second orientational orientational order parameter S_2 is a constructive one to the dielectric constant in the normal direction of monolayer films. The electro-optic Pockels effect can also be expressed and discussed in orientational orders. Discussing molecular motion and dielectric behavior of monolayers in orientational order parameters gives the systems information on polarity and $C_{\infty v}$ symmetry.

References

[1] J. N. Israelachvili, *Intermolecular and Surface Forces*, Academic, London (1985).

[2] C. X. Wu, Z. C. Ou-Yang, and M. Iwamoto, *J. Chem. Phys.*, **109** (1998) 4552.

[3] J. Topping, *Proc. R. Soc. London, Ser. A*, **114** (1927) 67.

[4] D. M. Taylor and G. F. Bayes, *Phys. Rev.*, **E49** (1994) 1439; R. E. Collin, *Field Theory of Guided Waves*, McGraw-Hill, New York (1960) Chap. 12.

[5] D. M. Taylor, O. N. Oliveira, Jr., and H. Morgan, *J. Colloid Interface Sci.*, **139** (1990) 508; J. R. MaCdonald and C. D. Barlow, Jr., *J. Chem. Phys.*, **39** (1963) 412; R. J. Demchak and T. J. Fort, Jr., *J. Colloid Interface Sci.*, **46** (1974) 191.

[6] M. Iwamoto, Y. Mizutani, and A. Sugimura, *Phys. Rev.*, **B54** (1996) 8186; C. X. Wu and M. Iwamoto, *Phys. Rev.*, **B55** (1997) 10922.

[7] A. Sugimura, M. Iwamoto, and Z. C. Ou-Yang, *Phys. Rev.*, **E50** (1994) 614.

[8] M. Iwamoto, C. X. Wu, and W. Y. Kim, *Phys. Rev.*, **B54** (1996) 8191.

[9] C. Kittel, *Introduction to Solid State Physics*, Wiley, New York (1974).

[10] C. X. Wu, W. Zhao, M. Iwamoto, and Z. C. Ou-Yang, *J. Chem. Phys.*, **112** (2000) 10548.

[11] T. L. Mazely and W. M. Hetherington III, *J. Chem. Phys.*, **86** (1987) 3640; J. D. Byers, H. I. Yee, T. Petralli-Mallow, and J. M. Hicks, *Phys. Rev.*, **B49** (1994) 14643; T. Petralli-Mallow, T. M. Wong, J. D. Byers, H. I. Yee, and J. M. Hicks, *J. Phys. Chem.*, **97** (1993) 1383.

[12] T. F. Heinz, in *Nonlinear Surface Electromagnetic Phenomena*, edited by H. E. Ponath and G. I. Stegemen, Elsevier Science, New York (1991) p.397-398.

[13] J. A. Giordmaine, *Phys. Rev.*, **138** (1965) 1599.

[14] Z. C. Ou-Yang and Y. Z. Xie, *Phys. Rev.*, **A32** (1985) 1189.

[15] Y. R. Shen, *The Principles of Nonlinear Optics*, Wiley, New York (1984).

[16] N. Bloembergen and P. S. Pershan, *Phys. Rev.*, **128** (1962) 606.

[17] H. Huang, G. Gu, S. Yang, J. Fu, G. K. L. Wong, and Y. Du, *Chem. Phys. Lett.*, **272** (1997) 427.

[18] K. D. Singer, M. G. Kuzyk, and J. E. Sohn, *J. Opt. Soc. Am.*, **B4** (1987) 968; J. D. Le Grange, M. G. Kuzyk, and K. D. Singer, *Mol. Cryst. Liq. Cryst.*, **150** (1987) 567.

[19] J. W. Wu, *J. Opt. Soc. Am.*, **B8** (1991) 142.

[20] M. Iwamoto, Y. Majima, H. Naruse, T. Noguchi, and H. Fuwa, *Nature*, **353** (1991) 645.

CHAPTER 3

MAXWELL DISPLACEMENT CURRENT METHOD

In the past few decades, the study of the electrical properties of the monolayer at the air -liquid interface (Langmuir film) has undergone dramatic developments through experiments, and some theoretical methods developed in recent years have promoted a more profound understanding of various kinds of monolayers at the air-liquid interface. [1] Such a broad interest was based on the specific electrical and optical characteristics due to the symmetry breaking at the surface, as discussed in section 1.4. [2] Various experimental techniques have been developed and have played a crucial role in this kind of study and allowed one to study the physico-chemical properties of monolayer films on a water surface without destroying the ultra-thin films. [3] One such significant technique is the Maxwell-displacement-current (MDC) measuring technique based on the idea illustrated in Fig. 1.3, [4] which allows the orientational change of polar molecules on a water surface to be probed by measuring MDC even in the range of immeasurably low surface pressures. The MDC is the transient current generated across the top and base electrodes separated by an air gap, e.g., due to the orientational change of polar molecules deposited on a water surface. Electrically, the MDC flowing across a monolayer on an air-liquid interface is caused by a change in the amount of induced charge on the top electrode suspended in the air as a result of monolayer compression. Thermostatistically, this is due to the change in the thermal average of the vertical component of the dipole moment of the constituent molecules of monolayers upon monolayer compression. MDC measurement can give essential information on the polar orientational order of monolayers in the entire range of molecular areas.

3.1 Maxwell Displacement Current

Any discussion of the dielectric properties of organic monolayers must include its polarization P which, under an external electric field E, is associated with the displacement field D as

$$D = \epsilon_0 E + P. \tag{3.1}$$

Here the polarization P is classified into spontaneous polarization P_0, first order or linear polarization, and nonlinear polarization. As will be seen in this chapter, Maxwell displacement current mainly comes from spontaneous polarization. Dielectric constant attributed to the first-order or linear polarization has been discussed in Chapter 2. The nonlinear polarization of organic monolayer films has been touched in Chapter 2 and will be discussed detailedly in Chapter 7.

According to the Maxwell's electromagnetic field theory, [5] the total current J flowing across an organic film is attributed to the conduction current and the displacement current, and is given by

$$J = env + \frac{\partial D}{\partial t}, \tag{3.2}$$

where e is the electron charge, n is the carrier density, and v is the velocity of the charge carriers. The first term of Eq. (3.2) represents the conduction current created by the movement of charge carriers. This current is the steady state current generated by the application of a potential difference between two electrodes which are separated by an organic film. Electrons injected from one electrode into the film are transported through it to the counter electrode under the external electric field formed between the two electrodes. Of course, this steady state current is very small in dielectric and electrically insulating films. In this sense, it is also called as the leakage current. However, the situation changes for organic ultra-thin films with a film thickness less than several nano-meters, where electrons may cross the films by tunneling. The control of electron tunneling opens a new way to electronics. For this purpose, techniques to prepare pin-hole free ultra-thin films must be developed. [6] Fortunately, several techniques have been developed to prepare excellent thin films during the last ten years. Using a novel preparation method based on the Langmuir-Blodgett (LB) technique, mono- and multilayer films of electrically insulating polyimide (PI) LB films have been successfully prepared. [7] The detail discussion concerning the electrical property of PI LB films and its application to electronics will be provided in Chapter 9.

The second term of Eq. (3.2) represents the Maxwell displacement current (MDC). It can be generated across the organic film due to the capacitance change such as the vibration of electrodes or due to the change of induced charges at the electrodes such as monolayers by the application of an external stimulus (for example, pressure, light, electric field or heat). In other words, the current irrelevant to free electronic movement belongs to displacement current. Sections 3.2, 3.3, and 3.4 are devoted to discussing the MDC by monolayer compression. Section 3.5 is about photoirradiated MDC and Section 3.6 is about one of its applications.

3.2 Maxwell Displacement Current by Monolayer Compression

As indicated in Chapter 2, organic monolayers have spontaneous polarization P_o due to the symmetry breaking at the interface, which is defined as Eq. (2.3) (see also Fig. 2.1). It is found from Eq. (3.1), Eq. (3.2), and Eq. (2.3) that the change of the order parameter S_1 contributes to the current flowing across the organic film. Such a characteristic can be made use of to detect the displacement current induced by monolayer compression on the water surface. The principle of MDC measurement for monolayers on the water surface is very simple as shown in Fig. 3.1. [8] One electrode (electrode 1) is suspended in the air and placed parallel to the water surface. The

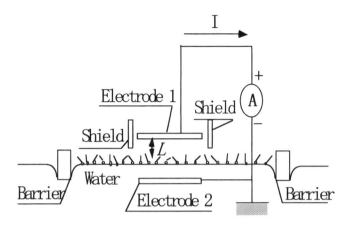

Figure 3.1 Schematic diagram of the experimental setup used in the present study.

other electrode (electrode 2) is immersed in the water subphase of the Langmuir trough. These two electrodes are connected to each other through an ammeter. The charge Q_1 induced on electrode 1 due to the polarized charge P_0 in the monolayers is given by [9]

$$Q_1 = -B\frac{\mu S_1}{LA} - C\Phi, \tag{3.3}$$

where B is the working area of electrode 1 L is the distance between the water surface and electrode 1, C is the capacitance between electrode 1 and the water surface, and Φ is the surface potential of water. During the course of monolayer compression and expansion at a compression speed of $\gamma = -dA/dt$, a displacement current I, as defined below,

$$I = \frac{B\gamma}{L}\frac{\mu S_1}{A^2} - \frac{B\gamma}{L}\frac{d(\mu S_1)}{AdA} - C\gamma\frac{d\Phi}{dA} \tag{3.4}$$

will be generated due to the change in the surface density of molecules under electrode 1 (first term), the change of the orientational order parameter S_1 (second term), and the change in the surface potential (third term). This means that MDC will not be generated without at least one of these changes. As we know, the charges induced is connected to current by an integratin performance of Eq. (3.4). Thus, firstly, the permanent dipole moment of the constituent molecules of monolayers and the order parameter can be determined from the MDC measurement. [10] Using this measurement, the dipole moment of various kinds of monolayers including fatty acids, mesogenic liquid-crystal monolayers have been determined. [9, 10, 11] Secondly, the phase transition of monolayers induced by monolayer compression can be detected, because MDC is very sensitive to the change of the order parameter S_1. [10, 12] Since the air-gap is a good electrical insulator, it eliminates the flow of leakage-current across the monolayer. Monolayers of fatty acids, cyano-biphenyl liquid crystal molecules and phospholipid molecules have been examined. It should be noted here that MDCs across monolayers can be measured even when a two-parallel electrode arrangement is not used. However, for the purpose of quantitative analysis of monolayers by MDC, it is best to choose the electrode arrangement system for which the analytical Laplace field solution is known, so that the charges induced on electrode on the basis of the Green's reciprocal theorem can be theoretically calculated. [9, 13, 14]

3.3 MDC Generated across Organic Monolayers Consisting of Molecules with Dielectric Anisotropy

The study of the two-dimensional alignment of molecules with two components of permanent dipole moments becomes very important for understanding the physico-chemical properties of monolayers due to their dielectric anisotropy. The relationship between the anisotropic dielectric constant and the permanent dipole moments in bulk liquid crystals is written as [15]

$$(\epsilon - I) \cdot E = (N/\epsilon_0)(\langle \alpha \cdot (E_i + E) \rangle + \langle \mu \rangle), \tag{3.5}$$

where N is the volume density of molecules, E is the external electric field, E_i is the internal electric field, and I is the unit matrix. It can be seen from Eq. (3.5) that the dielectric anisotropy $\Delta\epsilon$ ($= \epsilon_\parallel - \epsilon_\perp$) depends on the permanent moments, $\mu_\parallel$ and $\mu_\perp$, parallel and perpendicular to the molecular long axis, respectively. Here $\epsilon_\parallel$ and $\epsilon_\perp$ are the dielectric constants along and perpendicular to the molecular long axis, respectively. Böttcher calculated an anisotropic relation of liquid crystal in various bulk phases on the basis of Onsager theory. [16] In contrast, for monolayers at an air-water interface, the relationship given by Eq. (3.5) cannot be used for the analysis because of the anchoring effect of an ultra-thin monolayer on a water surface, and the electrostatic interaction occurring between polar molecules and a water surface. [17] In this chapter, a model is established for monolayer films consisting of biaxial molecules and measure the MDC of nCB (with permanent dipole along the molecular long-axis), DON103 (with larger permanent dipole perpendicular to the molecular long-axis), and LVI-035L (without permanent dipole) in the range of immeasurably low surface pressures. Subsequently, the dependence of the MDC on the anisotropy of the permanent dipole moment, paying attention to the generation of MDC in the range just after the phase transition from the planar surface alignment isotropic phase to the polar one upon monolayer compression is examined. It should be noted here that this kind of phase transition has been confirmed to occur in the authors' previous experimental and theoretical studies on mesogenic liquid crystals of nCBs. [18]

3.3.1 Molecular Structures and Experimental Results

To know the contribution of dielectric anisotropy, it is constructive to study three kinds of liquid crystal molecules, namely nCB (n = 5, 7, 8, and 10), DON103, and LVI-035L, the molecular structure of which is illustrated in Fig. 3.2, where the direction of the permanent dipole moment is indicated by arrows. For simplicity, each molecule is considered to be rodlike with its permanent dipole moment in the middle of the rod representing the average of the bond dipole moments distributed among the molecular chain. Although the direction of the C=O covalent bond is not perpendicular to the molecular long-axis in the actual DON molecule, it is assumed that the biaxial molecule is simplified as a rodlike axis with a length l, the smaller point dipole $\mu_{\parallel}$ along the molecular long-axis at the midpoint, and the larger dipole $\mu_{\perp}$ at a distance l_0 pointing perpendicular to the molecular long-axis because of the C=O covalent bond direction. This simplification does not diminish the physical base related to the MDC generation. The molecular weights of 5CB, 7CB, 8CB, 10CB, DON103, and LVI-035L are 251, 277, 290, 303, 317, and 276.5 respectively. nCB molecules have a positive dielectric anisotropy in nematic phase. For 5CB, it is estimated that $\Delta\epsilon(= \epsilon_{\parallel} - \epsilon_{\perp}) = 9.9$, [15] and its dipole moment is in the direction along the molecular long axis. DON103 has a negative dielectric anisotropy $\Delta\epsilon = -1.3$, and the larger permanent dipole points along CO bond perpendicular to the molecular long axis. It is a mixture of three types of DON with R/R' of C_3/C_2 (23%), C_4/C_5 (38%), and C_5/C_1 (39%), respectively. LVI-035L molecule is dielectric isotropic and possesses no total permanent dipole moment.

The experimental setup is exactly the same as Fig. 3.1. Briefly, the trough was rectangular with an area of 1,050 cm^2. At the center of the trough, electrode 1 with a working area of 44.2 cm^2 was suspended parallel to and at a distance of 1.0 ± 0.05 mm above the water surface. The two electrodes were connected to each other through a picoampere electrometer with high sensitivity, and MDCs generated across monolayers were measured as the monolayer was compressed or expanded with the aid of two movable barriers. Monolayers of nCB (n = 5, 7, 8, and 10), DON103, and LVI-035L were spread from a chloroform solution onto the water surface using a microsyringe. The monolayers on the water surface of the LB trough were compressed from both sides of the trough with two barriers moving simultaneously at a constant barrier velocity of 40 mm/min in opposite directions at room temperature.

Figure 3.3 shows the surface pressure-area isotherms of 5CB, DON103 and LVI-035L in the whole compression process, which were obtained by a Wilhelmy surface pressure measurement technique. [19] Upon monolayer compression, the surface pressure rose in a molecular area of around 150 Å^2 for DON103 and around 60 Å^2 for 5CB. For monolayers of LVI-035L, the surface pressure is very low over the entire range but it is not zero in a molecular area of around 170 Å^2. These experimental results reveal that monolayers were formed on a water surface. In the experiments, special attention is given to the generation of MDC in the range of immeasurably low surface pressures, i.e., in the range just after the phase transition from the planar

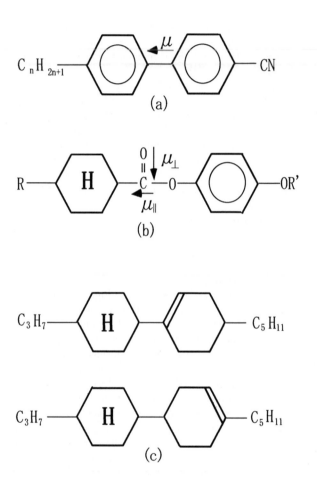

Figure 3.2 Molecular structures of (a) nCB, (b) DON103, and (c) LVI-035L.

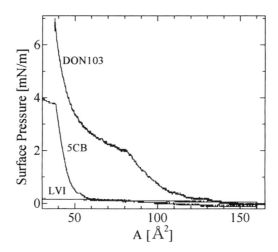

Figure 3.3 Experimental results for surface pressure in the whole compression region.

surface alignment isotropic phase to the polar orientational one.

The experimental results for MDC and surface pressure in the range of immeasurably low surface pressures are shown in Figs. 3.4, 3.5 and 3.6. As shown in the figures, in the region $A/A_0 > 1(A_0 = \pi l^2$, l is the length of the rodlike molecule), before the onset of the transition at the molecular area $A/A_0 = 1$, MDC I is zero. In contrast, after the onset of the transition, MDC I is about several fA in the range $0.8 < A/A_0 < 1.0$. The data obtained by MDC measurement was reproducible within the constraint of the experiment. [20] Here the investigation is constrained within the reorganization of monolayers in the range of immeasurably low surface pressures, where molecules lying on the water surface experience phase transition from the planar surface alignment isotropic phase to the polar one in the molecular area $A = A_0$ upon monolayer compression. The MDC of 5CB is provided as a typical example in the serial experiments on nCB, and it had been found that the MDCs of 7CB, 8CB, and 10CB have similar traces. As shown in Fig. 3.4, in the range of molecular areas $0.6 < A/A_0 < 1$, corresponding to $192 \text{ Å}^2 < A < A_0 = 320 \text{ Å}^2$, the MDC I of 5CB increased quickly to a value of several tens of fAs, which is larger than those of DON103 and LVI-035L. Further compression induced an increase in MDC to several hundred fAs (not shown in the figure). [17] As shown in Fig. 3.5, the MDC I

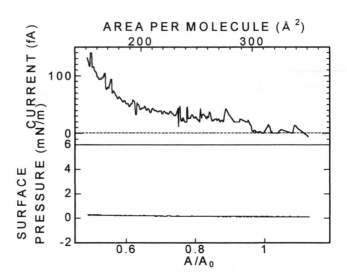

Figure 3.4 Maxwell displacement current of 5CB monolayers with a constant compression barrier speed of 40mm/min.

of DON103 is about ten fA in the range between 240 Å² and 300 Å² ($= A_0$). In the range $90 < A < 240$ Å², i.e., in the range $0.3 < A/A_0 < 0.8$, the displacement current remains less than 100 fA, whereas LVI-035L generated almost no MDC in the entire range of molecular areas upon monolayer compression, as shown in Fig. 3.6.

3.3.2 Molecule-surface Interaction

As described in some electrodynamics textbooks, [21] the generation of interaction energy $W(\theta)$ acting between polar molecules and the water surface, which is the first step in statistical calculation, can be simplified by assuming an image dipole $\{(\epsilon_w - 1)/(\epsilon_w + 1)\}\mu$ at a mirror-symmetric position against the water surface. Here ϵ_w is the relative dielectric constant of water and μ is the dipole moment of molecules. For half a membrane, $W(\theta)$ is calculated as [22]

$$W(\theta) = -\frac{\mu^2}{16\pi\epsilon_0 l^3 \cos\theta}\frac{\epsilon_w - 1}{\epsilon_w + 1}, \tag{3.6}$$

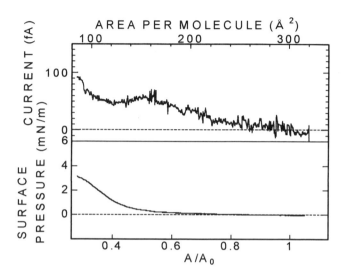

Figure 3.5 Maxwell displacement current of DON103 monolayers with a constant compression barrier speed of 40 mm/min.

where ϵ_0 is the permittivity of a vacuum, l is the length of the molecule along its axis, and θ is the tilt angle of hydrocarbon chains away from the normal direction of the monolayer. Despite the divergence of MDC at the molecular area $A = A_0$ by monolayer compression, which is mainly caused by the take-off of very small molecules lying on the water surface, the model of half a membrane reveals the common features quite well for monopolar-molecule monolayer such as 5CB, with a permanent dipole along the long axis of rodlike molecule. [22] However, the model of half a membrane cannot explain the present experimental results of MDC across monolayers consisting of biaxial molecules with a permanent dipole moment not parallel to the long molecular axis. For monolayers of cyclohexanecarboxylate-type liquid crystal (DON), with its larger permanent dipole perpendicular to the long molecular axis, the MDC generated by monolayer compression is rather small, around 10 fA [see Fig. 3.5] after the phase transition from the planar surface alignment phase to the polar one, in comparison with 100 fA for 5CB, which was reported by the authors. [17]

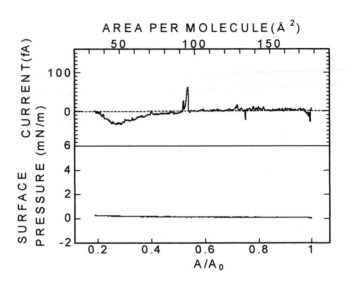

Figure 3.6 Maxwell displacement current of LVI-035L monolayers with a constant compression barrier speed of 40 mm/min.

3.3.3 Orientational Order Parameter for Biaxial Molecules

As S_1, defined in Eq. (1.2) [22, 23] is proportional to the spontaneous polarization normal to the monolayer surface[Eq. 2.3], it is one of the main contributors to MDC. For monopolar molecules with a permanent dipole parallel to the long molecular axis, either the Legendre polynomial definition or the physical definition given by Eq. (1.2) may be adopted, in the same manner as discussed by Sugimura *et al.* [22] However, for biaxial molecules with a permanent dipole not parallel to the long molecular axis, it is better to introduce two orientational parameters $S_\parallel$ and $S_\perp$, which are the order parameter of the dipole component along the molecular long axis and perpendicular to the long axis respectively. The total orientational order parameter S_1 and $S_\parallel$ and $S_\perp$ must satisfy the following relationship,

$$S_1 = (\mu_\parallel \langle \cos \theta_\parallel \rangle + \mu_\perp \langle \cos \theta_\perp \rangle)/\mu$$
$$= (\mu_\parallel S_\parallel + \mu_\perp S_\perp)/\mu, \tag{3.7}$$

where $\mu_\parallel$ and $\mu_\perp$ are the permanent dipoles along the long molecular axis and perpendicular to the molecular axis, respectively, $\theta_\parallel$ and $\theta_\perp$ are their corresponding tilt angles away from the normal direction of the monolayer (Fig. 3.7), and μ is the total dipole momentum of the molecule expressed as $\sqrt{\mu_\parallel{}^2 + \mu_\perp{}^2}$. From the expression above, it is obvious that S_1 is the degree of contribution by molecule dipoles in the normal direction with respect to the water surface on the basis of the statistical average. In this chapter, with a membrane model, this definition is used to discuss the contribution of polarization to the MDC of biaxial molecules such as DON on a water surface in the range of low surface pressure after the phase transition from the planar surface alignment phase to the polar one. Then this model is used to explain the experimental results for nCB (n = 5, 7, 8, and 10), DON103, and LVI-035, which differ from each other in the magnitude of the generation of MDC associated with the dielectric anisotropy due to molecular biaxiality. Finally the MDCs of these molecules are calculated on the basis of a biaxial molecule model expressible with two order parameters $S_\parallel$ and $S_\perp$, where $S_\parallel$ and $S_\perp$ are the orientational order parameter of these molecules in the directions parallel and perpendicular to the molecular long axis respectively. [17]

As described in the authors' paper, [17] the MDC I flowing across the electrometer is generated due to the orientational change of biaxial molecules upon monolayer compression. In the range of molecular areas $A > A_0$, the molecules are assumed to lie flat on the water surface owing to the electric attractive force acting between the molecules and the water surface, and consequently the vertical component of the dipole moment in the direction perpendicular to the water surface is zero. Thus the MDC is zero in this range (see Fig. 3.5). In the range of molecular areas $A < A_0$ after the transition in the molecular area A_0, molecules align on a water surface and MDC I generated by monolayer compression with constant monolayer compression speed $\gamma(= -dA/dt)$ given by Eq. (3.4) can be rewritten in another form

$$
\begin{aligned}
I &= I_\parallel + I_\perp \\
&= \frac{B\gamma}{L}\left(\frac{\mu_\parallel S_\parallel}{A^2} - \frac{d(\mu_\parallel S_\parallel)}{AdA}\right) + \frac{B\gamma}{L}\left(\frac{\mu_\perp S_\perp}{A^2} - \frac{d(\mu_\perp S_\perp)}{AdA}\right).
\end{aligned}
\tag{3.8}
$$

Here the effect of surface potential change of pure water is neglected, because the change is small in comparison with the effect of the change in the order parameter S caused by monolayer compression. [23] It should be noted here that all molecules lie flat on a water surface in the range of the planar surface alignment isotropic phase $(A > A_0)$, and they begin to "stand up" in the molecular area $A = A_0$ upon monolayer compression. Thus in the range of the polar orientational alignment phase $(A < A_0)$, molecules can align on a water surface. It is shown from Eq. (3.8) that the MDC is the sum of two terms, the first term due to the change in order parameter $S_\parallel$ along the molecular long-axis, and the second one due to that in the order parameter $S_\perp$ perpendicular to the molecular long-axis. Moreover, it can be calculated that the contribution of the axis-orienting dipole moment is much larger than that of the dipole moment perpendicular to the molecular long axis (Section 3.3.5). That is, the

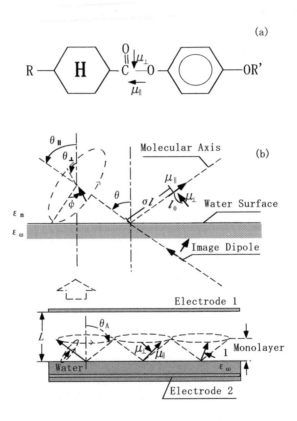

Figure 3.7 (a) Molecular structure of DON and (b) sketch of the model for the biaxial-molecule monolayer at an air-liquid interface.

first term dominates the generation of MDC. In the following section, discussion shall be confined to the MDC before and after the onset of the phase transition in the molecular area $A = A_0$.

3.3.4 Comparison between the Two Directions

Suppose a biaxial monolayer film with a permanent dipole not parallel to the molecular axis (e.g. DON) is spread on an air-water surface, as shown in the inset of Fig. 3.7 which illustrates the interaction of biaxial molecules with the water surface and where R and R' are carbon chains. Although the direction of the C=O covalent bond is not perpendicular to the long molecular axis in the actual DON molecule, it is assumed in the foregoing analysis that the biaxial molecule is simplified to a rodlike axis with a length l, with the smaller point dipole $\mu_\parallel$ along the long molecular axis at the midpoint, and the larger dipole $\mu_\perp$ at a distance l_0 perpendicular to the long molecular axis because of the C=O covalent bond direction. This model is different from the conventional treatment which was discussed by Sugimura *et al.*, [22] because it has a dipole moment $\mu_\perp$ perpendicular to the long molecular axis. The system experiences a three-degree-of-freedom thermodynamic motion, i.e., the motion of the rodlike long axis in the azimuthal plane, and the revolution and rotation of the long molecular axis shortly after the onset of the transition at the molecule area $A/A_0 = 1$ by monolayer compression. Here A is the area per molecule and A_0 is the critical molecular area estimated to be πl^2. The revolution gives no thermostatistical contribution to MDC because the orientational order parameter S_1 defined as Eq. (3.7) may not change, i.e., the vertical component of the dipole moment of molecules does not change. In contrast, the change in the distance of $\mu_\perp$ from the liquid surface while rotating, should lead to MDC generation. The rotation angular displacement φ and tilt angles $\theta_\parallel$ and $\theta_\perp$ at an arbitrary position, as shown in Fig. 3.7, satisfy

$$\cos \theta_\perp = \cos \varphi \sin \theta \tag{3.9}$$

and

$$\theta_\parallel = \theta. \tag{3.10}$$

The molecules are assumed to lie on the water surface at $A > A_0$ and align in the range of $0 \leq \theta \leq \theta_A$ due to the effect of hardcore intermolecular forces in the range of molecular area $A < A_0$, where $\theta_A = \arcsin(\sqrt{A/A_0})$. Therefore, in the range $A > A_0$, the order parameter S_1 defined by Eq. (3.7) is zero (Fig. 3.7) and MDC I is not generated by monolayer compression. The discussion is confined to monolayers in the range $A < A_0$. At first, the interaction between $\mu_\perp$ and $\mu_\parallel$ is neglected because of its independence in a molecule. Furthermore, the interaction among molecules can be neglected in the region $0.6 < A/A_0 < 1$, because the interaction energy among molecules $W_d = 11.0342\mu^2/4\pi\epsilon_0 a^3$ (μ the permanent dipole moment, a the distance between the most neighboring two molecules) [18] is smaller than the interaction energy between water and molecule W_s given by Eq. (3.11) in this range, under the assumption that $a = 4\sigma l \sin \theta_A$ ($0 < \sigma < 1$). Here σl is the distance from the water surface.

For simplicity, discussion shall focus on the generation of MDC considering only the interaction between molecules and the water surface, because the main interaction is the generation of MDC shortly after the onset of transition at the molecular area $A/A_0 = 1$. By introducing an image dipole $\{(\epsilon_w - 1)/(\epsilon_w + 1)\}\mu$ at a mirror position with respect to the water surface (see Fig. 3.7), the effect of interface interaction working between a dipole with a permanent dipole moment μ and the water surface is obtained as

$$W(\theta) = -\frac{\mu^2(\cos^2\theta + 1)}{64\pi\epsilon_0 d^3}\frac{\epsilon_w - 1}{\epsilon_w + 1}. \tag{3.11}$$

Here d is the distance between the dipole and the water surface. For $\mu_\perp$, d satisfies

$$d = \sigma l \cos\theta - l_0 \cos\varphi \sin\theta, \tag{3.12}$$

where σl is the distance between the point dipole $\mu_\parallel$ and the water surface along the long molecular axis. Under the approximation of

$$d^{-3} \approx (\sigma l \cos\theta)^{-3}(1 + 3\frac{l_0}{\sigma l}\tan\theta\cos\varphi) \tag{3.13}$$

and using Eqs. (3.9) and (3.11),the interaction energy $W(\theta_\perp)$ of $\mu_\perp$ can be written as

$$W_\perp(\theta) = -\chi_\perp kT\frac{\sin^2\theta\cos^2\varphi + 1}{\cos^3\theta}(1 + 3\frac{l_0}{\sigma l}\tan\theta\cos\varphi) \tag{3.14}$$

with

$$\chi_\perp = \frac{\mu_\perp^2}{8\pi\epsilon_0 l^3 kT}\frac{\epsilon_w - 1}{\epsilon_w + 1}. \tag{3.15}$$

The limit of the approximation Eq. (3.13) must be estimated because $l_0\tan\theta/(\sigma l) = \infty$ in the case of $\theta = \pi/2$. Taking DON as a typical example, $\mu_\perp = 2.5$ D, $\sigma l = 1.0$ nm [15], and $l_0 = \mu_\perp/(2e) = 0.025$ nm are used, and $\chi_\perp = 9.2 \times 10^{-3}$ and $\theta < 76°$ if $l_0\tan\theta/(\sigma l) < 0.1$ are obtained. That is, the calculation based on Eq. (3.13) is reasonable within the region $[0°, 76°]$ with a deviation under 10%. On the other hand, substituting

$$d = \sigma l \cos\theta_\parallel \tag{3.16}$$

and Eq. (3.10) into Eq. (3.11) leads to

$$W_\parallel(\theta) = -\chi_\parallel kT\frac{\cos^2\theta + 1}{\cos^3\theta}, \tag{3.17}$$

where

$$\chi_\parallel = \frac{\mu_\parallel^2}{8\pi\epsilon_0 l^3 kT}\frac{\epsilon_w - 1}{\epsilon_w + 1}. \tag{3.18}$$

For DON, a typical value of $\mu_\parallel = 0.8$ D leads to $\chi_\parallel = 1.0 \times 10^{-3}$. The monolayer is treated as a boson system which satisfies the Boltzmann distribution

$$f(\theta, \varphi) = \frac{e^{-W/kT}}{Z}, \tag{3.19}$$

where $W = W_\perp + W_\parallel$. As in the region $[0°, 76°]$, $W_\perp/kT \ll 1$ and $W_\parallel/kT \ll 1$, the single-particle partition is

$$Z = \int_0^{\theta_A} \int_0^{2\pi} e^{-(W_\perp + W_\parallel)/kT} \, d\varphi \sin\theta d\theta$$
$$= 2\pi[(1 - \cos\theta_A) + Z_\perp + Z_\parallel], \qquad (3.20)$$

where

$$Z_\perp = \int_0^{\theta_A} (-\frac{W_\perp}{kT}) \sin\theta d\theta$$
$$= (\frac{\ln\cos\theta_A}{2} - \frac{3\cos^2\theta_A - 1}{4} \frac{}{\cos^2\theta_A})\chi_\perp \qquad (3.21)$$

and

$$Z_\parallel = \int_0^{\theta_A} (-\frac{W_\parallel}{kT}) \sin\theta d\theta$$
$$= (-\ln\cos\theta_A + \frac{1 - \cos^2\theta_A}{2\cos^2\theta_A})\chi_\parallel. \qquad (3.22)$$

$Z_\perp$ and $Z_\parallel$ are the rectification to the free space, which is expressed by $2\pi(1 - \cos\theta_A)$. With the Boltzmann distribution defined as Eq. (3.19), the orientational order of $\mu_\perp$ is given by

$$S_\perp = \int_0^{\theta_A} \int_0^{2\pi} \sin\theta \cos\varphi \frac{e^{-(W_\perp + W_\parallel)/kT}}{Z} \, d\varphi \sin\theta d\theta. \qquad (3.23)$$

A simple calculation of Eq. (3.23) under the approximation $e^{-(W_\parallel + W_\perp)/kT} \approx 1 - (W_\parallel + W_\perp)/kT$ leads to

$$S_\perp(\theta_A) = \frac{1}{Z} \frac{3\pi\chi_\perp l_0}{4l} (\frac{32}{3} + \frac{7}{3\cos^3\theta_A} - \frac{10}{\cos\theta_A} - 3\cos\theta_A), \qquad (3.24)$$

where Z is also a function of θ_A expressed by Eq. (3.21). From Eq. (3.24), it can be seen that the order parameter of $\mu_\perp$ is small because of a small $\chi_\perp$. The numerical $S_\perp - A/A_0$ relation of DON is presented in Fig. 3.8(a), which reveals that $S_\perp$ undergoes a sharp decrease to zero as the monolayer is compressed, indicating that $S_\parallel$ is the main contributor to MDC shortly after the onset of transition. It should be noted here that $S_\perp$ is plotted as the solid line in the range $A/A_0 < 0.94147$ because of the calculation limitation of $0° < \theta_A < 76°$. The orientational order of the dipole $\mu_\perp$ perpendicular to the long molecular axis is very small, around a scale of 0.01, and quickly reduces to zero shortly after the onset of the transition. It is obvious that when $A/A_0 \to 0$, i.e., $\theta \to 0$, $S_\perp \to 0$, which corresponds to the state that $\mu_\perp$ is parallel to the water surface. This result can also be easily obtained from Eq. (3.24). On the other hand, in the case of $\mu_\parallel$, the order parameter $S_\parallel$ should be

$$S_\parallel = \frac{\int_0^{\theta_A} \int_0^{2\pi} e^{-(W_\parallel + W_\perp)/kT} \cos\theta \sin\theta d\varphi d\theta}{Z}. \qquad (3.25)$$

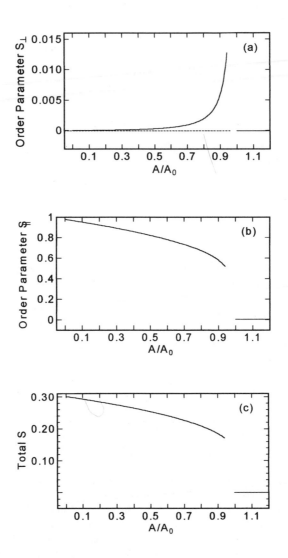

Figure 3.8 (a) Order parameter $S_\perp$ with respect to relative molecular area A/A_0. (b) Relative area dependence of order parameter $S_\parallel$. (c) Total parameter S_1 exhibits the same shape as $S_\parallel$.

A similar calculation of Eq. (3.25) yields

$$S_\parallel = \frac{(1 - \cos^2 \theta_A) + (\cos \theta_A - \frac{1}{\cos \theta_A})(2\chi_\parallel + \chi_\perp)}{2[(1 - \cos \theta_A) + Z_\perp + Z_\parallel]}, \tag{3.26}$$

which is also depicted in Fig. 3.8(b). The orientational order of the dipole $\mu_\parallel$ along the molecular axis presents a similar shape as obtained by Sugimura et al. with the half-membrane treatment. [22] The total orientational order S_1 defined by Eq. (3.7) is presented in Fig. 3.8(c). The total MDC I generated by monolayer compression with the constant speed $\gamma = -dA/dt$ is given by

$$\begin{aligned} I &= I_\perp + I_\parallel \\ &= \frac{B\gamma}{L}\left(\frac{\mu_\perp S_\perp + \mu_\parallel S_\parallel}{A^2} - \frac{d(\mu_\perp S_\perp + \mu_\parallel S_\parallel)}{AdA}\right). \end{aligned} \tag{3.27}$$

Here B is the working area of the electrode, and L is the distance between the water surface and the electrode suspended in air above the water surface in the measurement of MDC (electrode 1 in Fig. 3.1). It is found that Eq. (3.27) is nothing but the displacement current generated by an apparent polarization of $P_z = \mu_\perp S_\perp + \mu_\parallel S_\parallel$. Due to the extremely complicated expression for I, its explicit form is not given but only shown by a graph (Fig. 3.9).

So far, the orientational order of the molecules with a permanent dipole moment not parallel to the long molecular axis and MDC in the region $[0°, 76°]$, i.e., in the region $0 < A/A_0 < 0.94147$ has been calculated. For better comparison between $\mu_\perp$ and $\mu_\parallel$ in the contribution of MDC, two ratios are designated:

$$\alpha_S = \frac{\mu_\perp S_\perp}{\mu_\parallel S_\parallel} \tag{3.28}$$

and

$$\alpha_I = \frac{I_\perp}{I_\parallel}, \tag{3.29}$$

which are shown in Fig. 3.10. Figure 3.10(b) also reveals that the direction of MDC created by $\mu_\perp$ is different from that by $\mu_\parallel$, and that the total orientational order S_1 and MDC of biaxial molecules are mainly created due to the dipole moment $\mu_\parallel$ parallel to the long molecular axis despite its much smaller dipole moment in comparison with the dipole moment $\mu_\perp$ perpendicular to the long molecular axis, as the monolayer is compressed. This conclusion is also supported by the finding that the total orientational order S_1 (Fig. 3.8(c)) defined by Eq. (3.7) has the same appearance as $S_\parallel$ (Fig. 3.8(b)). These results coincide with the fact that molecules with different main chain lengths exhibit the onset of the MDC generation at the molecular area $A_0 = \pi l^2$, in a manner similar to the generation of nCBs. [18] Obviously, liquid crystal monolayers on a water surface exhibit various phases during monolayer compression, and the interaction among molecules becomes increasingly important as monolayers

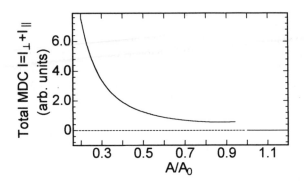

Figure 3.9 Theoretical total MDC I.

are further compressed. [18] However, in the model, only the electrostatic interaction between molecules and the water surface is taken into account. Thus it should be noted here that the present model is useful only in the range of low surface pressure, just after the phase transition at the molecular area A_0 from the planar alignment phase to the polar one, because the dipole moment parallel to the long molecular axis is the dominant contributor to displacement current. It is also obvious that the MDC of 5CB generated due to a much larger permanent dipole along the long molecular axis should be much larger than that of DON. The MDC generation of DON103 shown in Fig. 3.5 is very small and it flows in the positive direction. The shape of the MDC is similar to that shown in Fig. 3.9 obtained by calculation. Taking DON103 as an example, the magnitude of MDC obtained by simple calculation in the range of $0.8 < A/A_0 < 0.94147$ is several fA, which is within the same scale as the experimental results (Fig. 3.5). Here it should be noted that the calculation above is restricted to within $[0°, 76°]$. At the beginning of the "standing-up" of DON103 molecules, i.e., in the region $0.94147 < A/A_0 < 1$, the deviation of calculated values obtained in this work becomes bigger. On the other hand, as $A/A_0 \to 1$, the O atom in the C=O bond still rests on the monolayer because the C=O bond is more hydrophilic than the C-C bond. Thus at this stage the monolayer is half a membrane and MDC is mainly created by $\mu_\perp$ in a manner similar to that described by Sugimura *et al.*'s calculation. [22] The reason why there was not such a large peak at the onset of

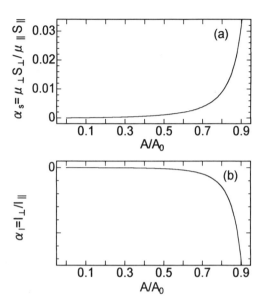

Figure 3.10 Comparison between the two dipoles $\mu_\perp$ and $\mu_\parallel$. (a) α_S shows that the effect of $\mu_\perp$ is negligible shortly after the onset of the transition. (b) Ratio α_I shows the opposite MDC direction and a great discrepancy between the contributions of $\mu_\perp$ and $\mu_\parallel$.

transition as that predicted by the half-membrane model is that at $\theta = \pi/2$, the size of the molecule should be considered and the interaction energy (Eq. (3.6)) should be rectified into some other form in which the convergence at A/A_0 is avoided. As has been discussed in this chapter, molecules are assumed to lie and stand up on a water surface and not to be pulled into the water. For further detailed discussion, this effect must be taken into consideration.

It is now possible to explain the compression process. Immediately after the onset of the transition at the molecular area $A/A_0 = 1$, $\mu_\perp$ is the main contributor to MDC, but its contribution decreases exponentially as the molecule area is further compressed. In a very short time, $\mu_\parallel$ replaces $\mu_\perp$ and dominates the contribution to MDC in the compression process thereafter.

3.3.5 Simulation

From Fig. 3.10, it is found that the directions of the MDCs created by $P_\perp$ and $P_\parallel$

are opposite. Moreover, the MDC of DON103 molecules is mainly created by the dipole along the molecular long axis despite its much smaller dipole moment during monolayer compression. This leads to the conclusion that the orientational change of the dipole moment along the molecular long axis, i.e., the dipole moment along the main chain, dominates the MDC generation. Thus it is obvious that the experimental MDC of nCB should be much larger than those of DON103 and LVI-035L after the onset of transition in the molecular area $A = A_0$, as nCB have a positive anisotropy and their axis-orienting permanent dipole moments are much larger than those of DON103 and LVI-035L (3.6 D for nCB compared with 0.8 D for DON103 and 0 D for LVI-035L). [24] The MDCs generated due to $\mu_\perp$ and $\mu_\parallel$ in the monolayer flow in opposite directions, as shown in Fig. 3.10 for DON103 simulation. However, the MDC generated due to $\mu_\perp$ is completely canceled out owing to the positive MDC generation of $\mu_\parallel$, as seen in Fig. 3.10(b). As for LVI-035L, because there is no total permanent dipole moment, no MDC generation was observed [see Fig. 3.6].

In order to compare the extents of MDC generation in nCB, DON103 and LVI-035L, MDC of nCB, DON103 and LVI-035L by Eq. (3.8) are also calculated and the results are plotted (Fig. 3.11(a)). The permanent dipole moments of nCB were chosen as $\mu_\parallel = 3.6$ D, and $\mu_\perp = 0$ D, [24] and those of LVI-035L were chosen as $\mu_\parallel$ and $\mu_\perp = 0$, on the basis of the MDC experimental results. [17] The value of l is chosen as 1 nm in the calculation for nCB, as that in the calculation for DON103 because of almost the same experimental critical area A_0 (see Figs. 3.4 and 3.5), and χ is taken to be 0.153. In Fig. 3.11(b), the experimental results obtained by the MDC measurement are also plotted. The MDC generation in nCB is much more extensive than that in DON103 because the permanent dipole moment of nCB is much larger than that of DON103. As can be seen in the figure, the experimental results shown in Fig. 3.11(b) and the theoretical calculation results shown in Fig. 3.11(a) show similar behavior with monolayer compression, though there is a discrepancy in the magnitude of MDC. This might be due to the assumption of zero dipole moment in the direction perpendicular to the molecular long axis for nCB. The divergence at the onset of transition ($A/A_0 = 1$) is attributed to the negligence of the size of molecules, which would lead to the divergence of Coulomb interaction at A/A_0 when the distance between dipoles and their images is zero, and the dipole-dipole interaction as expressed by Eq. (3.11) becomes inapplicable. Therefore, the simulation range was chosen as $\theta_A < 0.94147$ so that the deviation would be less than 10%. For detailed understanding of the MDC at the transition, for example, calculation in which the size and shape of molecules are considered needs to be carried out. Right after the onset of the transition, $\mu_\perp$ of DON103 contributed to the MDC generation somehow, but quickly decreased to zero because of the almost zero thermal average along the z direction of rotational motion. The MDC in LVI-035L is zero according to the calculation results in this chapter. The agreement between theoretical treatment and experimental results indicates that the simple biaxial molecule model of monolayers with dielectric anisotropy can explain the generation process of MDC just after the phase transition on the basis of a simple

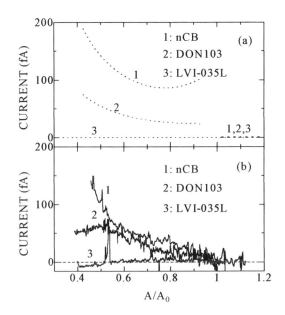

Figure 3.11 (a) Theoretical simulation of MDC in nCB, DON103 and LVI-035L. (b) Experimental results for MDC in nCB, DON103 and LVI-035L.

model of the transition from the planar surface alignment isotropic phase to the polar one, taking into account the orientational distribution of the constituent polar molecules, which is ruled by Boltzmann statistics. This treatment will be of help for detailed understanding of the physico-chemical properties of monolayers. However in a real monolayer system, the organization of monolayers on a water surface upon monolayer compression may not be so simple, possibly because of the existence of islands.

3.4 Phase Transition of Chiral Phospholipid Monolayers by Maxwell Displacement Current Measurement

3.4.1 Maxwell Displacement Current of Chiral Phospholipid Monolayers

Phospholipids (PCL) are one of main amphiphiles of biomembranes and they contain one phosphatidylcholine (polar hydrophilic) head group and two long alkyl chains with a carbonyl (hydrophobic) group (Fig. 3.12). Usually, the hydrophilic and hy-

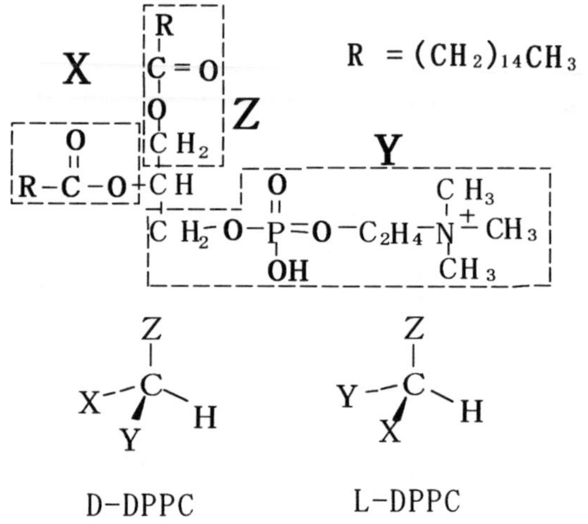

Figure 3.12 Schematic diagram for DPPC molecule.

drophobic groups are not located in the same plane containing the two hydrocarbon chains. Thus the PCL molecules are referred to as chiral due to the lack of the mirror plane symmetry. Traditionally D-DPPC is defined to have right-handedness and L-DPPC left-handedness. Another remarkable physico-chemical property of the amphiphiles is that they can show a variety of structures in the bulk of aqueous solutions, and their thermodynamics and intra-aggregate forces in bulk of solutions have been extensively investigated. [25] Unfortunately, the role of the chirality of the amphiphiles in the structure multilicity has not been studied clearly. Since monolayers at the air-water interface show some different phases during monolayer compression and can reveal more information on the molecular conformation than in bulk, it is important to examine the molecular conformation of chiral and racemic compounds in the monolayer state. For the similar purpose, Lanquist [26] found that for monolayers of 2-alkanos and certain derivative racemates, pure enantiomers form different crystal-like phases. The studies of chiral and racemic 12-hydroxyoctadecanoic acids (12HOA) later showed that the molecules at the water-air surface can take the two conformations with straight chain and bent chain, respectively, while the packing molecular models are proposed with rectangular and oblique lattices for racemic and chiral 12HOAs, respectively. [27] The same dependence of both kinds of lattice depending on the chirality has been discovered in an amphiphilic monolayer. [28] Although the measurement of surface pressure-area $(\pi - A)$ isotherm is performed in the monolayers

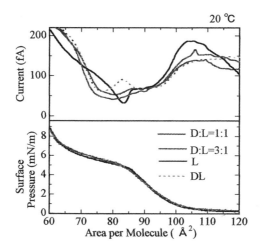

Figure 3.13 $\pi - A$ isotherms (below) and corresponding Maxwell displacement currents (above) measured in four DPPC monolayers composed of pure L-DPPC, DL-DPPC, and mixtures with molar ratios of D-:L-DPPC=1:1, 3:1, respectively.

at water-air interface, the above measurement of molecular packing and conformation is carried out by transferring the monolayers onto mica [28] or calcium fluoride plate [27] using the Langmuir Blodgett technique. [29] This process may cause a change in both molecular packing and conformation affected by the substrate-specific. Therefore, it is a challenge to detect the molecular conformation in monolayer at the air-water surface without transfer so that the monolayers are not destroyed.

Figure 3.13 shows the typical experimental results of the measurement for four samples: pure L-DPPC and DL-DPPC monolayers and their mixtures with a molar ratio of D-:L-DPPC=1:1, 3:1, respectively. (The experimental setup is shown in Fig. 3.1.) The monolayers of mixtures of D-, L-, and DL-DPPC in designed ratios are formed on the water surface by spreading their dilute chloroform solutions onto the water surface of the Langmuir trough. The monolayers of DPPC formed on the water surface were compressed with two floating barriers at a constant barrier velocity of 40 mm/min, i.e., at a compression speed of 0.0081 Å^2/sec. The MDC and $\pi - A$ isotherms were simultaneously measured during the monolayer compression. As shown in Fig. 3.13, the $\pi - A$ isotherms of these four samples exhibit almost identical behavior relative to molecular area A containing nearly the same pressure plateau in the range of 65 to 85 Å^2 of molecular area at a temperature of 20°C. The latter is usually referred to as the two-dimensional (2D) phase transition region from the expanded monolayer to condensed one. With the help of the X-ray diffraction, Eckhardt argued that the 2D positional transition associates with the molecular conformation change from bent chain state to straight chain state. In the authors' recent work, the plateau region is interpreted to involve the phase transition of molecular orientation from isotropic to polar orientation. [22] From the present $\pi - A$ measurement it seems that the mentioned phase transitions are obviously independent of the chirality of the molecules composed in the monolayers. However, the results of measurement in MDC shown in Fig. 3.13 above reveal that although the main shapes of the MDC viewed as a function of the molecular area A are roughly similar for the four samples of DPPC, there exist some essential differences between them. The most striking characteristic is the generation of MDC peak which appears in the beginning of the pressure-plateau, $A = 85$ Å^2, with different amplitude and sign for the four examples. From the viewpoint of molecular level, the four examples of the DPPC monolayers differ only from their chirality: DL-DPPC is racemic and L-DPPC and D-DPPC are both chiral with different optical activity. Therefore, the mentioned MDC peak should reveal some molecular conformation change relating to the molecular chirality. In other words, this anomalous MDC generation may be of significant help to understanding the chiral discrimination from molecular conformation level. Due greatly to the problem being of critical importance in physics, chemistry, biology and geology, [30] a theory has been presented to describe the chirality dependence of the MDC peak of a monolayer, and discuss the anomalous MDC generation. [31]

3.4.2 Symmetry of MDC Generation Influenced by Chirality in Mixed Phospholipid Monolayers

In order to examine the polarization dependence on molar ratio, MDC generation from pure chiral L- and D-DPPC monolayers and mixtures are detected, where the most striking characteristic was the generation of anomalous MDC peak appearing in the beginning of the pressure-plateau. From Fig. 3.13, it is found that the mixed monolayers exhibit a W shape of MDC in the region around 70-90 Å^2. A parameter Δ_1/Δ_2 is defined to describe the feature of the W-shape MDC in the beginning of the pressure-plateau, as shown in Fig. 3.14(a). Figure 3.14(b) shows the dependence of the depth ratio between the two valleys of the W shape for pure L- and D-DPPC monolayers, and mixtures with a molar ratio of L-:D-DPPC=1:2, 1:1, and 2:1. From Fig. 3.14(b), it is found that the MDC feature in this region shows a chirality symmetry between left- and right- handedness, which well supports the theoretical explanation given by the authors. [32] In other words, the anomalous MDC generation due to the chiral twist conformation transition depends on the mixing ratio, with the molar ratio 1:1 as its minimum point. This result indicates that the chirality of molecules influences the MDC generation, which means that the chirality discussed here is a nontrivial new order.

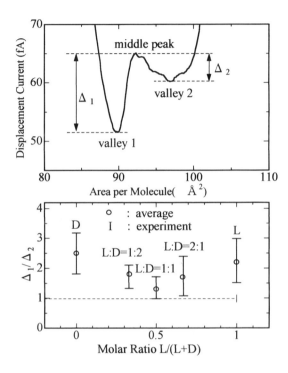

Figure 3.14 (a) Maxwell displacement current feature with a W shape for DPPC monolayers. (b) Feature of the W-shape MDC for pure L- and D-DPPC, and their mixtures with molar ratios of L-:D-DPPC=1:2, 1:1, and 2:1, respectively.

3.5 Maxwell Displacement Current by Photoirradiation

The activity of molecules in response to photoirradiation always receive scientists' attention. [33] For example, the photochromic properties of azobenzene derivatives, such as AZBPAA shown in Fig. 3.15, have been established for the development of optical memories: trans-to-cis isomerization occurs when they are irradiated with ultraviolet light, and the reverse occurs at visible wavelengths (see). [34] The detection

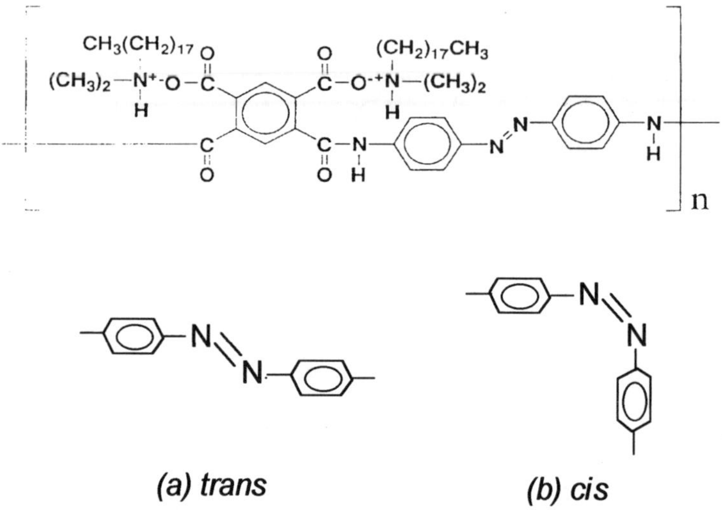

Figure 3.15 Molecular structure of AZBPAA.

of the photoelectric response of single monolayers is very important for a better understanding of the relationship between the structure and function of monolayers. The MDC technique has proved to be a good candidate for this purpose.

Figure 3.16 shows the experimental arrangement. Electrode 1, composed of a transparent SnO_2-coated glass slide, is suspended in air, parallel to the water surface, whereas electrode 2 is immersed in the water surface. The two electrodes are connected by a gold wire through a sensitive electrometer, whose internal electrical resistance is zero. A 500-W xenon lamp is used to induce photoisomerization in a single monolayer. The ultraviolet and visible light outputs of the lamp are separated using a glass filter. The azobenzene used is AZBPAA with a molecular structure shown in Fig. 3.15. The azobenzene is delivered onto the surface of deionized water from a benzene solution to form a single monolayer. Most of the AZBPAA molecules are supposed to be in the *trans* form when they are spread on the water surface. The generation of MDC across precursor monolayers of polyimide containing azobenzene derivatives (AZBPAA) due to photoisomerization is examined by the electrometer with high sensitivity and high accuracy by irradiating the monolayers through trans-

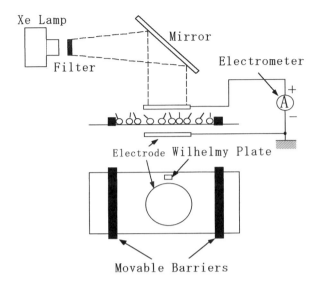

Figure 3.16 Experimental setup of MDC measurement by photoirradiation.

parent electrode suspended in the air (see Fig. 3.16). [35, 36]

Figure 3.17 shows a typical example of the MDC obtained by alternating irradiation with ultraviolet light (λ_1) with a wavelength of 380 nm and visible light (λ_2) with a wavelength of 450 nm (repeated twice), followed by irradiation with visible light with a wavelength of 450 nm using the experimental setup (Fig. 3.16). The alternating irradiation generated a MDC in the negative direction when ultraviolet light irradiation (λ_1) was applied and in the opposite direction when visible light (λ_2) was applied. The total charge flowing through the circuit during the ultraviolet light irradiation was almost the same as that flowing during the visible light irradiation. These results indicate that the configurational change of the AZBPAA molecules induced by the cis-trans photoisomerization is reversible; that is, the change in the vertical component of dipole moment given by μS_1 due to the cis-trans photoisomerization is reversible. On the other hand, irradiation twice in succession with visible (or ultraviolet) light does not generate MDC. Similar experimental results were obtained for other monolayers containing azobenzene derivatives. The direction of MDC generated with alternating photoirradiation depends on the chemical structure of azobenzene monolayers, the molecular area and so on.

The generation of MDC from monolayers deposited on a conducting electrode

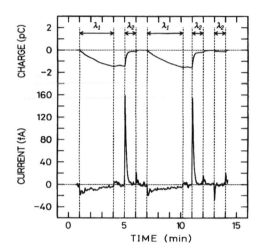

Figure 3.17 Generation of MDC from a single monolayer of AZBPAA on a water surface by irradiations with UV (λ_1) and visible light (λ_2). The MDC charge flowing through the circuit is also shown. The distance between electrode 1 and the water surface was 1.07 mm.

can also be examined. [35, 36] Suppose an air gap between two conducting electrode is separated by a good electrical insulator such as polytetrafuloroethylene (PTFE) sheet at a spacing of several ten micro-meters. The charge induced on the counter electrode is given by the first term of Eq. (3.3). For monolayers on solid substrate, there is no possibility that the MDC current is generated due to a change in the number of molecules; that is, a main contribution to the MDC generation is the change in the vertical component of dipole moment of molecules, m_z ($= \mu S_1$), caused by the photo-induced cis-trans photoisomerization. For AZBPAA monolayers deposited on a glass slide, the displacement current is generated in a manner similar to that generated on the water surface as shown in Figure 3.18,[35, 36] whereas it is not generated in devices without monolayers. It is interesting to compare the charge flowing across monolayers on solid substrates during the photoirradiation with the charge flowing across monolayers on the water surface during the photoirradiation. Figure 3.19 shows plots of the charge $\Delta Q/C$(open circles, $C = C_1$) generated from a single monolayer on a water surface and of the charge $\Delta Q/C$(closed circles, $C = C_2$) generated from a single monolayer on a solid substrate during irradiation with visible light. Here, C_1 is the capacitance formed between suspended electrode and the water surface (see

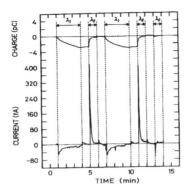

Figure 3.18 Generation of MDC from a AZBPAA monolayer on ITO glass slide by irradiation with UV (λ_1) and visible light (λ_2). A cell with a structure of a base-ITO electrode/monolayer/ air-gap/top-ITO electrode was used. The distance between two electrodes was about 10 μm The MDC charge flowing through the circuit is also shown.

Fig. 3.16), and C_2 is the device capacitance. On water and on solid substrates, the irradiation-induced changes in the vertical component of average dipole moment of the constituent monolayer molecules seem to be almost the same. Therefore, the MDC is generated across AZBPAA monolyers on the water surface and on a solid substrate in a similar manner.

However, generally, the generation of MDC depends on monolayer-film conditions. For example, the MDC across monolayers of fatty acids containing azobenzene on solid substrate is different from that on a water surface. [37, 38]

3.6 Evaluation of Liquid Crystal Alignment Using MDC Technique

As described in section 3.5, MDC is very sensitive to the change of molecular motion of monolayers. For monolayers on solid substrate, the induced charge comes only from the first term of Eq. (3.3), which can be applied to detect the change of polarization P_0 for monolayers. Due to the configurational change of the constituent molecules, phase transition of monolayers and so on, MDC shows the characteristic behaviors of monolayers applicable in electronics. For example, as mentioned in previous section, azobenzene monolayers experience a cis-trans configurational change by the alternating photoirradiation of ultra-violet (UV) and visible light, and results

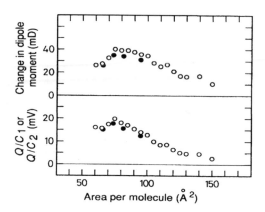

Figure 3.19 Plots of the charge $\Delta Q/C_1$ (open circles) generated from a single monolayer on a water surface and of the charge $\Delta Q/C_2$ (closed circles) generated from a single monolayer on SnO_2 coated glass slide, as a function of the area per molecule. Change in the vertical component of dipole moment (Δm_z) is also plotted.

in the alternating change in the polarization P_0. From the viewpoint of application such as memory storage media, this alternating change is very important because it enables one to distinguish two configurational forms of molecules by the direction of MDC. Similar arguments will be held for the cases when other external stimuli such as pressure, heating etc [39] are used.

Recently, much attention has been paid to the photoregulation of liquid crystal alignment. [40] Among these issues of interest is the orientational control of liquid crystal (LC) alignment triggered by the photoisomerization of surface azobenzene layers on the substrate. [41, 42] The MDC measurement has been applied to the investigation of monolayers of poly(vinyl alcohol)s bearing azobenzene side chains (6Az5PVA as shown in Fig. 3.20) mixed with p-pentyl-p'-cyano-biphenyl (5CB) on a water surface. [43] For mixed monolayers containing trans-form 6Az5PVA and 5CB, MDC was not generated by alternating photoirradiation of UV and visible light. In contrast, for mixed monolayers containing cis-form 6Az5PVA and 5CB, MDC was detected (see Fig. 3.21). Of interest is that 6Az5PVA monolayers deposited in cis-form can photoregulate the alignment of 5CB molecules, while 6Az5PVA monolayers deposited in trans-form can not.

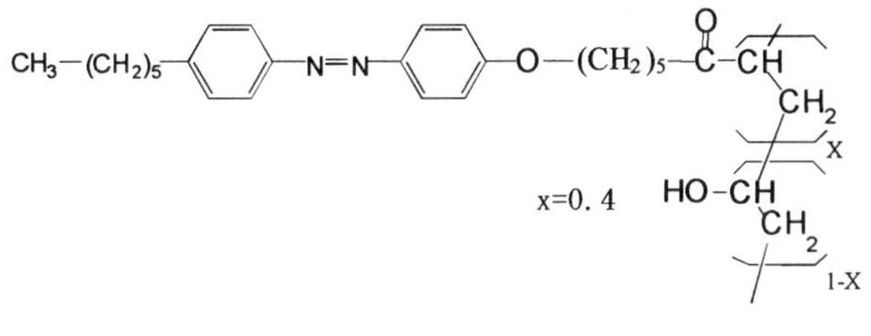

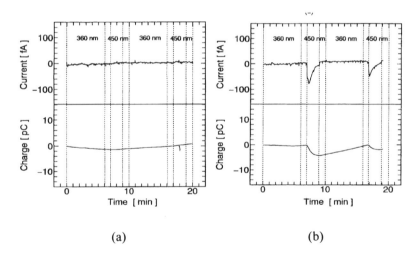

Figure 3.20 Molecular structure of 6Az5PVA.

Figure 3.21 A typical example of photoinduced MDCs across mixed monolayers on a water surface. (a) trans-6Az5PVA and 5CBs. (b) cis-6Az5PVA and 5CBs.

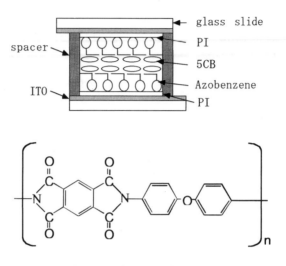

Figure 3.22 NLC cells with surface azobenzene monolayer and chemical structure of PI. (a) LC cells and (b) PI.

The orientational transition process of 5CB molecules in nematic LC (NLC) cells triggered by the configurational change of the surface 6Az5PVA azobenzene mono-layer on PI LB films has been examined by measuring the optical transmittance and the electrical capacitance at the same time. Figure 3.22 shows the LC cells and the chemical structure of PI LB. PI LB films are deposited onto indium-tin-oxide (ITO)-coated glass slides by means of the precursor method in combination with the vertical dipping method in the LB technique, at a surface pressure of 35 mN/m. [44] [Briefly, the process of preparing ultrathin PI LB films consists of three steps. First, mono-layer films of poly(amic acid) long alkyl amine salts, the precursor of PI, are spread onto a pure water surface (pH 5.8); Second, the poly(amic acid) salt monolayer films are transferred onto a solid substrate by raising or lowering the substrate through the water surface; And finally, PI multilayer films are fabricated of poly(amic acid) salt multilayer films by chemically treating poly(amic acid) films with a mixture of acetic anhydride and pyridine. The number of deposited layers is controlled by changing the number of dipping cycles. After the deposition of PI LB films, 1-layer trans-form azobenzene polymer, poly(vinyl alcohol)s bearing azobenzene side chains (6Az5PVA) (see Fig. 3.20), are deposited at a surface pressure of 30 mN/m from a pure water surface. Two types of LC cells can be prepared. For the first type, both top and bottom ITO -coated glass slides are covered with PI LB films and a 1-layer 6Az5PVA monolayer (denoted as Type I). For the second type, the top ITO-coated glass slide is only covered with PI LB films while the bottom ITO -coated glass slide is covered with PI LB films and a 1-layer 6Az5PVA monolayer (denoted as Type II).] The top and bottom glass slides were attached face to face with the dipping direction of the two glass slides antiparallel to each other at a spacing of about 14 μm, using a supporting adhesive epoxy resin incorporating beads with a diameter of 14 μm. The working area B of LC cells is around 1.50 cm^2. The cell gap L was measured by an optical interference method, and was determined to be 14.0 ± 0.5 μm. The capacitance of the cell was measured before and after filling the LC cells with nematic p-pentyl-p'-cyano-biphenyl (5CB) liquid crystals. The 5CBs have a positive dielectric anisotropy $\Delta \epsilon$ ($=\epsilon_\parallel$ -$\epsilon_\perp$) of 10.5 and a birefringence Δn($=$ n$_\parallel$ -n$_\perp$) of 0.182, where $\epsilon_\parallel$($ =16.3$) and $\epsilon_\perp$($=5.8$) are the relative dielectric permittivities parallel and perpendicular to the long molecular axis, and n$_\parallel$ ($=1.709$) and n$_\perp$($=1.527$) are the reflective indices parallel and perpendicular to the long molecular axis, respectively. [45] For 5CBs on 3-, 5-, 11-layer PI LB films, the pretilt angles are estimated by an optical method to be 0.46°, 0.31°, and 0.93°. For the sake of simplicity, the pretilt angle of 5CBs on PI LB films was assumed to be nearly zero for the analysis. Figure 3.23 gives the schematic of the experimental arrangement of the simultaneous capacitance and transmittance measurement during the alternating photoisomerization of UV and visible lights or during the application of a step voltage. The two electrodes were connected to an LCZ meter. A power supply was connected to an external voltage input terminal of the LCZ meter. The capacitance of the LC cell was measured with the LCZ meter at the frequency of 1 kHz. The oscillation level was set at 50 mV, which is well below the optical threshold of the cell. A dc step voltage was added onto the ac voltage to the LC cell by using the power supply, when it was required. (It takes 0.1 sec for getting one value of capacitance using the LCZ meter and a computer.) The transmittance of a He-Ne laser beam (632.8 nm in wavelength) through the LC cells placed between two crossed polarizers was measured during the alternating photoisomerization of UV and visible lights or during the application of a step voltage. It should be noted that the He-Ne laser beam does not affect the photoisomerization of the azobenzene chromophore under the experimental condition. The angle between the dipping direction and the polarized light through the polarizer was 45 degree. The light intensity of a He-Ne laser beam through polarizer, cell, analyzer, and bandpass filter was measured by using an electrometer and a Si-photo-diode. The transient of the light intensity is monitored with a digitizing oscilloscope or a digital multimeter.

Figure 3.24 gives the experimental results of the capacitance and the optical trans-

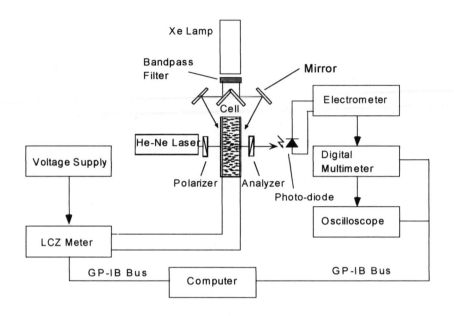

Figure 3.23 Experimental setup of LC cell measurement.

mittance for such LC cells with 0, 1, 5, 11 layers of PI LB films. Of interest is that
the transmittance and the capacitance of the LC cells with 0- or 1-layer PI LB films
are changed reversibly by the alternating irradiation of ultraviolet (UV) and visible
light, whereas those of LC cells with more than 5-layer PI LB layers are not changed
reversibly. [46] In Fig. 3.24, θ_0^- represents the tilt angle of 5CB molecules in LC cells.
These results are explained by taking into account the difference in the anchoring
strength of PI LB films, i.e., the anchoring strength of PI LB films with 5 and 11
layers is stronger than those with 1 layer and 0 layer. [47]

It is believed that the motion of 5CB molecules triggered by the surface azobenzene monolayer is ruled by the continuum theory describing the free energy of LC
cells and the anchoring energy at the interface. [46] During the alternating photoirradiation with UV and visible light, the surface azobenzene monolayer undergo cis-trans
photoisomerization, and as a result, the easy axis direction changes. Such an interfacial conformational change activates the deformation of bulk NLCs, by which the
directors undergo redistribution so as to satisfy the new boundary condition. The

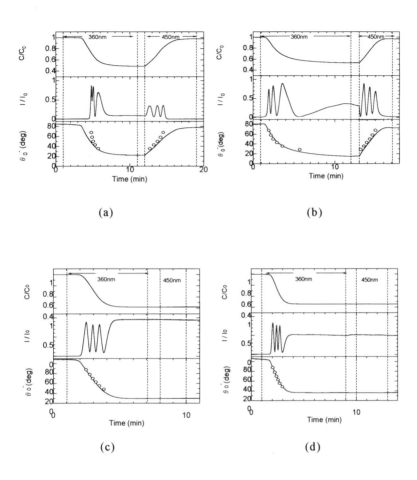

Figure 3.24 Typical example of capacitance (C/C_0) and optical transmittance (I/I_0) by alternating irradiation of UV (360 nm in wavelength) and visible light (450 nm in wavelength). Tilt angles θ_0^- estimated from capacitance (C/C_0) are plotted by solid line, and those estimated from the optical transmittance are plotted by open circles. (a) Without PI LB film layers, (b) with 1-layer PI LB film layer, (c) with 5-layer PI LB film layer, and (d) with 11-layer PI LB film layer.

capacitance of LC cells depending on the director orientation distribution is given by

$$C = C_0 \frac{1}{\frac{1}{L} \int_0^L \frac{dz}{\epsilon_\| \sin^2 \theta(z) + \epsilon_\perp \cos^2 \theta(z)}} \qquad (3.30)$$

where $C_0 = \epsilon_0 B / L$. Similarly, the intensity of the transmitted He-Ne laser beam through the LC cells is also a function of the director orientation distribution:

$$I = I_0 \sin^2 \left(\frac{\pi}{\lambda} L \Delta \right) \qquad (3.31)$$

with

$$\Delta = \frac{1}{L} \int_0^L \left\{ \frac{n_\| n_\perp}{\sqrt{n_\|^2 \sin^2 \theta(z) + n_\perp^2 \cos^2 \theta(z)}} - n_\perp \right\} dz,$$

in which the experimental arrangement shown in Fig. 3.23 has been considered. Here, λ represents the wavelength of the He-Ne-laser and I_0 represents the intensity of the input laser beam.

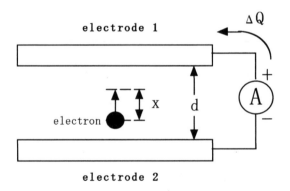

Figure 3.25 MDC generated by electron transfer.

For molecular electronic device application, the motion of charged particles must be detected and purposefully controlled in the planned and artificial mono- and multi-layer systems. The generation of MDC due to the electron transfer in multi-layer systems is helpful, because electron hops in the artificially arranged multilayer systems which are excited by laser-pulse can be detected as MDC pulses. [48, 49] Here the transfer of the number of N_t electrons at a distance x in the direction normal to the surface of two parallel electrodes gives rise to the change of induced charge ΔQ on the electrode. ΔQ is given by

$$\Delta Q = -eN_t x/d. \tag{3.32}$$

Here d is the distance between the two electrodes and $eN_t x$ corresponds to the change of polarized charge P_0 given by Eq. (2.3) (see Fig. 3.25).

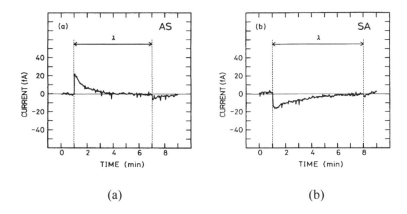

(a)   (b)

Figure 3.26 (a) Photoinduced MDC generated from $SnO_2/PIBM/A/S/air\text{-}gap/SnO_2$ junctions. (b) A typical example of displacement current generated from $SnO_2/PIBM/S/A/air\text{-}gap/SnO_2$ junctions.

In order to clarify the electron transfer, samples with structures of $SnO_2/PIBM/A/S/air\text{-}gap/SnO_2$ and $SnO_2/PIBM/S/A/air\text{-}gap/SnO_2$ are examined. [50] Here, A and S are single monolayers of PI and of PI containing tetraphenyl porphyrin moiety (PORPI). They work as electron acceptor and photosensitizer, respectively. Polyisobutylmethacrylate (PIBM)) multilayer films are good electrical insulators. Photoexcited electrons are expected to transfer between adjacent single monolayers A and S in a direction from S to A by photoirradiation, and the MDC pulses are found to be generated in opposite directions for the above two structures, as shown in Fig. 3.26.

In the MDC measurement, the response time depends on the time constant of the closed circuit. By measuring the photoinduced potential, the evaluation of the response time have been carried out and a value of the order of nano second has been recorded. [48, 49] Similar experiments have also been carried out for a condenser-type acceptor/sensitizer/donor hetero LB film to realize photogenerated charge storage. [51]

Here it should be noted that any dynamical motion of charged particles makes a contribution to the displacement current generation. Thus, it is very important to clarify the origin of displacement current generation in mono- and multilayer systems for the realization of purposeful electron transfer between the two adjacent single monolayers.

3.7 Summary

In this chapter, the dependence of MDC on the permanent dipole moment orientation in the range of immeasurably low surface pressures by means of an MDC measurement technique was examined. It is found that the MDC in a monolayer on a water surface depends mainly on the orientational change of the axis-orienting permanent dipole moment. The experimental results of the chiral DPPC monolayers shows that MDC also depends on molecular chirality. Photoizomerization in monolayers on a water surface as well as on a solid substrate can be detected by MDC technique. Likewise, MDC is also a good measurement technique to observe the liquid crystal alignment and to detect the electron transfer.

References

[1] G. Roberts, *Langmuir-Blodgett Films*, Plenum, New York (1991).

[2] G.L, Gaines, *Insoluble Monolayers at Liquid-Gas Interface*, Interscience, New York (1965).

[3] A. Ulman, *Characterization of Organic Thin Films*, Butterworth-Heinemann, Boston (1995).

[4] M. Iwamoto and Y. Majima, *J. Chem. Phys.*, **94** (1991) 5135.

[5] R. E. Collin, *Field Theory of Guided Waves*, McGraw-Hill, New York (1960) Chap.1.

[6] A. Ulman, *Ultrathin Organic Thin Films*, Academic Press, San Diego (1991).

[7] M. Iwamoto and M. Kakimoto, *Polyimides as Langmuir-Blodgett Films*, in *Polyimides, Fundamentals and Applications*, ed. Malay K. Ghosh and K. L. Mittal, Marcel Dekker, Inc., New York (1996) Chap. 25.

[8] M. Iwamoto, *Thin Solid Films*, **244** (1994) 1031, and references cited therein.

[9] M. Iwamoto and Y. Majima, *J. Chem. Phys.*, **94** (1991) 5135.

[10] M. Iwamoto, T. Kubota, and M. R. Muhamad, *J. Chem. Phys.*, **102** (1995) 9368.

[11] Y. Majima and M. Iwamoto, *Rev. Sci. Instrumn.*, **62** (1991) 2228.

[12] M. Iwamoto, T. Kubota, and Z. C. Ou-Yang, *J. Chem. Phys.*, **104** (1996) 736.

[13] M. Iwamoto, M. Fukuzawa, and T. Hino, IEEE Trans. Electr. Insul., **EI-22** (1987) 419.

[14] P. M. Morse and H. Feshbach, *Methods of Theoretical Physics*, McGraw-Hill, New York (1953).

[15] W. H. De Jeu, *Physical Properties of Liquid Crystalline Materials*, Gordon and Breach, London (1980).

[16] C. J. F. Böttcher, *Theory of Electric Polarization*, Elsevier, Amsterdam (1973), Vol. I.

[17] C. X. Wu and M. Iwamoto, *Jpn. J. Appl. Phys.*, **36** (1997) 824.

[18] M. Iwamoto, Y. Mizutani, and A. Sugimura, *Phys. Rev. B.*, **54** (1996) 8186; M. Iwamoto, T. Kubota, and M. R. Muhamad, *J. Chem. Phys.*, **102** (1994) 9368.

[19] L. Wilhelmy, *Ann. Phys.*, **119** (1863) 177.

[20] M. Iwamoto, Y. Majima, H. Naruse, T. Noguchi, and H. Fuwa, *Nature*, **353** (1991) 645.

[21] C. Kittel, *Introduction to Solid State Physics*, Wiley, New York (1974); J. N. Israelachvili, *Intermolecular and Surface Forces*, Academic Press, New York (1985).

[22] A. Sugimura, M. Iwamoto, and Z. C. Ou-Yang, *Phys. Rev.*, **E50** (1994) 614.

[23] Y. Majima, A. Watanabe, and M. Iwamoto, *Jpn. J. Appl. Phys.*, **30** (1991) 126.

[24] Y. Nitta, Denki Zetzuen Zairyo-no Kagaku, *Chemistry of Electrical Insulating Materials*, Baifukan, Tokyo (1983) [in Japanese].

[25] N. Israelachvili, *Intermolecular and Surface Forces*, Academic, London (1992) Chaps. 16-18.

[26] M. Lundquist, *Arkiv. Kem.*, **17** (1960) 183.

[27] T. Tachibana, T. Yoshizumi, and K. Hori, *Bull. Chem. Soc. Jpn.*, **52** (1979) 34.

[28] C. J. Eckhardt, N. M. Peachey, D. R. Swanson, J. M. Takacs, M. A. Khan, X. Gong, J. H. Kim, J. Wang, and U. A. Uphans, *Nature*, **362** (1993) 614.

[29] K. Blodgett, *J. Am. Chem. Soc.*, **56** (1935) 1007.

[30] A. Collet, M. J. Brienne, and J. Jacques, *Chem. Rev.*, **80** (1980) 215.

[31] Z. C. Ou-Yang, X. B. Xu, C. X. Wu, and M. Iwamoto, *Phys. Rev.*, **E59** (1999) 2105; W. Zhao, C. X. Wu, Z. C. Ou-Yang, and M. Iwamoto, *J. Chem. Phys.*, **110** (1999) 12131.

[32] M. Iwamoto, C. X. Wu, and W. Zhao, *J. Chem. Phys.*, (2000)

[33] H. Durr and H. Bouas-Laurent, *Photochromism Molecules and Systems, Studies in Organic Chemistry 40*, Elsevier, Amsterdam (1990).

[34] Z. F. Liu, K. Hashimoto, and A. Fujishima, *Nature*, **347** (1990) 658.

[35] M. Iwamoto, Y. Majima, H. Naruse, T. Noguchi, and H. Fuwa, *Nature (London)*, **353** (1991) 645.

[36] M. Iwamoto, Y. Majima, H. Naruse, T. Noguchi, and H. Fuwa, *J. Chem. Phys.*, **95** (1991) 645.

[37] Y. Majima, Y. Kanai, and M. Iwamoto, *J. Appl. Phys.*, **72** (1992) 1637.

[38] M. Iwamoto, Y. Majima, and H. Naruse, *J. Appl. Phys.*, **72** (1992) 1631.

[39] M. Iwamoto, C. X. Wu, and W. Y. Kim, *Phys. Rev.*, **B54** (1996) 8186.

[40] A. A. Sonin, *The Surface Physics of Liquid Crystals*, Gordon Breach Publishers, Amsterdam (1995).

[41] K. Ichimura, Y. Suzuki, T. Seki, A. Hosoki and K. Aoki, *Langmuir*, **4** (1988) 1214.

[42] T. Seki, T. Tamaki, Y. Suzuki, Y. Kawanishi, K. Ichimura, and K. Aoki, *Macromolecules*, **22** (1989) 3505.

[43] W. Y. Kim, M. Iwamoto, and K. Ichimura, *Jpn. J. Appl. Phys.*, **35** (1996) 5395.

[44] M. Iwamoto, W. Y. Kim, and C. X. Wu, *J. Chem. Phys.*, **106** (1997) 9815.

[45] A. Sugimura, T. Miyamoto, M. Tsuji, and M. Kuze, *Appl. Phys. Lett.*, **72** (1998) 329.

[46] M. Iwamoto, K. Kato, A. Matsumura, and Y.Majima, *Jpn. J. Appl. Phys.*, **38** (1999) 5984.

[47] A. Sugimura, K. Matsumoto, Z. C. Ou-Yang, and M. Iwamoto, *Phys. Rev.*, **E54** (1996) 5217.

[48] M. C. Petty, *Langmuir-Blodgett films*, Cambridge, New York (1996).

[49] E. G. Wilson, *Electron. Lett.*, **19** (1983) 237; E. G. Wilson, *Jpn. J. Appl. Phys.*, **34** (1995) 3775, and references cited therein; X. B. Xu, T. Kubota, and M. Iwamoto, *Jpn. J. Appl. Phys.*, **35** (1996) 3630.

[50] M. Iwamoto, Y. Majima, M. Atsuzawa, M. Kakimoto and Y. Imai, *Phys. Rev.*, **B46** (1992) 10479.

[51] K. Naito, A. Miura, and M. Azuma, *Thin Solid Films*, **210/211** (1992) 268.

CHAPTER 4

MONOLAYERS VIEWED AS POLAR LIQUID CRYSTALS

Investigations of smectic-C liquid crystals (LCs) can be traced back to the 1970s and most of them are based on the elastic theory. [1] An elastic theory for smectic-C liquid crystalline systems has already been given by de Gennes et $al.$ [2, 3, 4] However, many difficulties were encountered in developing a theory for the orientational smectic-C–smectic-A phase transition. Apart from the reports by Gieβelmann and co-workers, [5, 6] scientific papers concerning the polar orientational smectic-A–smectic-C phase transition for liquid crystals are very rare. At the same time, some excellent experimental observations of the orientational phase transition between smectic-A and smectic-C LCs in monolayers on a water surface have been achieved by monolayer compression. [7, 8] For ferroelectric LCs, the electromagnetic interactions appear to be inevitably associated with this kind of transition. This is more obvious for polar monolayers, as for them the dipole-dipole interaction is very important in this case. [9, 10] Although some of the theoretical results for liquid crystals are applicable to the analysis of monolayers, an alternative theory for single monolayers is expected due to the dimensional difference between them (two dimensions for monolayers rather than three dimensions for liquid crystals) and the change in molecular states as a result of being subjected to external stimulation, such as monolayer compression. Moreover, monolayer films are structurally different from liquid crystals. The molecular area is invariable in liquid crystals, and the LC systems have central symmetry ($D_{\infty h}$ symmetry). In contrast, the molecular area of monolayer films varies in the monolayer compression process and normal-director monolayers have polar symmetry (C_{∞} symmetry), because of the interface spatial restriction. In addition, the in-plane molecular configuration may influence the phase transition in monolayer films through the internal electric field. Cai and Rice constructed a general density function theory to show the phase transition from a hexagonal non-tilted to a distorted hexagonal tilted structure. [11, 12] In monolayer films, two kinds of phase transition, i.e., orientational phase transition and configurational phase transition, are possibly involved because of their possible in-plane structural change (change of molecular configuration), which is different from that of LCs. That the positional configuration of molecules will influence the intermolecular interaction should be considered in such kind of phase transition study. In other words, the development of a more general description of this phase transition including the orientational influence and configurational influence becomes important.

Recently there has been an upsurge of interest in the study of the relationship between the monolayer structure and the dynamic features of monolayer systems. [10, 13, 14] The monolayer films have been found to be closely similar to biological membrane and have been used to explore the physical properties of membrane structure. These systems often represent the principal component of biological membrane in two manners. First, these monolayers can be viewed as two-dimensional (2D) systems determined by geometrical configuration on the molecule scale. Second, the orientational state of the constituent rodlike molecules in monolayers determines the dielectric properties in the direction normal to the monolayer surface. The state of monolayers usually can be characterized by certain order parameters. [9] In monolayers of polar molecules on a water surface, there are at least two order parameters: one is the molecular configuration, i.e., the positional distribution pattern of the heads of the molecules on a water surface, [15, 16, 17] and the other is the orientational distribution of the molecular tails. [9] These two order parameters have been found to be the fundamental parameters for monolayer films and can be connected to some physical quantities via some mathematical implementations. The molecular configuration in the monolayer plane plays an important part in many physical phenomena concerning Langmuir-Blodgett films. [10, 15, 16] The aim of this chapter is to provide a sound physical picture to describe the behavior of the polar orientational phase transition in smectic monolayers that are structurally different from bulk LCs on a water surface under monolayer compression. A self-consistent polar orientational phase transition equation for smectic monolayers with a consideration of the molecular configuration is derived and a detailed explanation of the polar orientational phase transition is offered. Finally the change of the first orientational order parameter and the differential dielectric constant at this phase transition point are examined.

4.1 Model and Internal Electric Fields

Consider a tilted-director smectic phase with a tilt angle θ_c as schematically shown in Fig. 4.1. The coordinate system is chosen in such a way that the smectic monolayer planes are parallel to the xy plane and the monolayer normal falls along the positive z axis. Constituent molecules of monolayers are assumed to be rodlike and have a permanent dipole moment μ in the direction parallel to the molecular long axis. The angle that the dipole moment at the origin makes with the layer normal is denoted by θ. As the monolayer concerned is amphiphilic, the hydrophilic parts, the head groups of the dipoles in the monolayer, are pinned onto the water surface due to the strong hydrophobic interaction, while the tails of these dipoles are free to rotate in the air. Thus, in the case of monolayers, the dipoles prefer to line up in the same direction above the water surface rather than take an anti-ferroelectric state in LCs, due to the surface component of the interaction. The dipole at the origin discussed is assumed to be restricted within the angular range $[0, \theta_A]$ with $\sin^2 \theta_A = A/\pi l^2$ (A is the molecular area, and l is the length of the long molecular axis), [9] due to the hardcore interaction among molecules in monolayer films. A strong steric repulsion takes places if θ approaches θ_A. The hardcore interaction produces no internal electric field in monolayer films, and can be regarded as an infinitely high potential well between 0

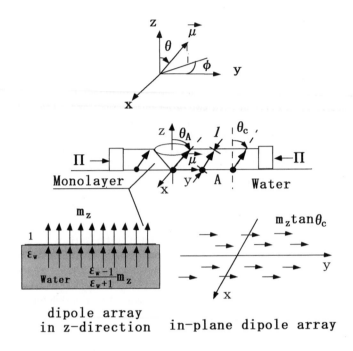

Figure 4.1 Basic geometry of the molecular-orientational model.

and θ_A for the tilt angle θ. Here it should be noted that as parts of the molecules may be more or less immersed in the water, only the effective part above the water surface is considered, possibly because the part immersed in the water is electrically screened compared with the part protruding in the air. The molecular area decreases as the monolayer films are compressed. The electric field produced by the tilted dipole array is decomposed into two directions: one standing perpendicular on the water surface and the other projection of the tilted dipole array onto the smectic monolayer plane. For convenience, the coordinate system is selected in such a way that the in-plane projection dipole array on the monolayer surface orients in the positive y direction, so that the internal electric field can be written as $\overrightarrow{E} = (0, E_y, E_z)$. The electric field at the origin is the sum of the fields created by the dipole array (excluding the dipole at the origin), and it is given by [18, 19]

$$E_y = m_z \tan \theta_c \cdot \frac{2\epsilon_w}{\epsilon_w + 1} \cdot \sum_{dipoles} \frac{3(e_y \cdot r_i)r_i - r_i^2 e_y}{4\pi\epsilon_0 r_i^5} \cdot e_y$$

$$E_z = -m_z \frac{2\epsilon_w}{\epsilon_w + 1} \cdot \sum_{dipoles} \frac{1}{4\pi\epsilon_0 r_i^3}, \tag{4.1}$$

where m_z is the depolarization of the monolayer in the z direction, ϵ_w is the relative dielectric constant of the material surface, for example, the water surface (77.6 at 25°C), ϵ_0 is the dielectric permittivity of a vacuum, r_i is the spatial vector of the molecules in the monolayer plane, and e_y is the unit vectors along the y axis. Here the factor $2\epsilon_w/(\epsilon_w + 1)$ $[= 1 + (\epsilon_w - 1)/(\epsilon_w + 1)]$ is added considering the effective dipole array m_z and an image dipole array with a dipole moment $(\epsilon_w - 1)m_z/(\epsilon_w + 1)$ induced at the $z = 0$ interface because of the interaction between the dipolar molecules and the water surface (see Fig. 3.1). The summation in Eq. (4.1) is carried over all sites of dipoles except the origin of the infinite 2D dipole array. For further treatment, it is convenient to introduce Euler angles for the dipole at the origin $\mu/\mu = (\sin\theta\sin\phi, \sin\theta\cos\phi, \cos\theta)$. Here ϕ is the azimuthal angle of the dipole discussed.

4.2 Polar Orientational Phase Transition in Smectic Monolayers

As have been verified by some excellent experimental observations, [7, 8] the smectic monolayer films undergo a polar orientational phase transition from a tilted-director phase to a normal-director phase as the monolayer films are compressed. The interaction energy in the tilted-director smectic phase can be considered as the sum of the interaction energy due to the dipole component array perpendicular to the monolayer film and the interaction energy due to the projection of the tilted dipole array onto the smectic monolayer plane. It is written as [18]

$$W_t(\theta, \phi) = -(g_y \sin\theta_c \sin\theta \cos\phi - g_z \cos\theta_c \cos\theta)a^{-3/2}\mu^2, \tag{4.2}$$

where g_y and g_z are defined as

$$g_y = 2a^{3/2} \sum_{dipoles} \frac{3(e_y \cdot r_i)r_i - r_i^2 e_y}{4\pi\epsilon_0 r_i^5} \cdot e_y$$

$$g_z = 2a^{3/2} \sum_{dipoles} \frac{1}{4\pi\epsilon_0 r_i^3}. \tag{4.3}$$

g_z is the same as Eq. (2.21) and $a = A/A_0$ is the relative molecular area (A and A_0 have been defined in Chapter 3). In a first-order approximation, $m_z = \mu\cos\theta_c$ is set in Eq. (4.2). The definitions as written by Eq. (4.3) are different from the definition by the authors before [18] by a factor of μ^2/kT and are based on the assumption $\epsilon_w \gg 1$. Eqs. (4.2) reveals that the component dipole array normal to the monolayer surface and the in-plane projection dipole array play different roles for the monolayer. The electric field produced by the component dipole array normal to the monolayer surface tends to maintain the normal-director smectic phase, while the electric field produced by the in-plane projection dipole array provides the force to deviate the constituent molecules from the normal direction. The competition between these two forces leads to a stable tilted-director smectic phase. The self-consistent theory requires that in the smectic phase the average director deviation or the y-axis orientational order parameter $\langle \sin\theta \cos\phi \rangle$ should be equal to θ_c, the tilt angle of the director of the monolayer film. That is

$$\langle \sin\theta \cos\phi \rangle \approx \theta_c \tag{4.4}$$

under small director deviation approximation. As it is assumed that the orientational distribution of the constituent molecules in the monolayer is ruled by Boltzmann statistics, the y-axis orientational order parameter is then given by

$$\langle \sin\theta \cos\phi \rangle = \frac{1}{Z} \int_0^{2\pi} \int_0^{\theta_A} \sin\theta \cos\phi e^{(g_y \sin\theta_c \sin\theta \cos\phi - g_z \cos\theta_c \cos\theta)a^{-3/2}\mu^2/kT} \sin\theta d\theta d\phi$$
$$\equiv f(\theta_c), \tag{4.5}$$

where Z is the single-partition function given by

$$Z = \int_0^{\theta_A} \int_0^{2\pi} \exp\{-W_t(\theta,\phi)/kT\} d\phi \sin\theta d\theta. \tag{4.6}$$

The integration of Eq. (4.5) is not unreasonably difficult. In an attempt to express the complicated integration of Eq. (4.5), two parameters $p = g_y \sin\theta_c a^{-3/2}\mu^2/kT$ and $q = g_z \cos\theta_c a^{-3/2}\mu^2/kT$ are introduced, and the Bessel function $I_\nu(z)$ which satisfies [20]

$$\frac{d}{dz}[z^\nu I_\nu(z)] = z^\nu I_{\nu-1}(z)$$
$$\frac{2\nu}{z}I_\nu = I_{\nu-1} - I_{\nu+1}. \tag{4.7}$$

Special cases of Bessel functions when $\nu = 0$, 1, and 2 are

$$I_0(z) = \frac{1}{\pi} \int_0^\pi e^{z\cos\phi} d\phi$$
$$I_1(z) = \frac{1}{\pi} \int_0^\pi e^{z\cos\phi} \cos\phi d\phi$$
$$I_2(z) = \frac{1}{\pi} \int_0^\pi e^{z\cos\phi} d\phi - \frac{2}{\pi z} \int_0^\pi e^{z\cos\phi} \cos\phi d\phi. \tag{4.8}$$

Using the definition of Bessel function Eq. (4.8) will significantly simplify the representation of the y-axis orientational order parameter Eq. (21). The integration with respect to ϕ of Eq. (4.5) is in fact of the same form as $I_1(z)$, and the y-axis orientational order parameter $\langle \sin \theta \cos \phi \rangle$ [Eq. (4.5)] becomes

$$\langle \sin \theta \cos \phi \rangle = \frac{2\pi e^{-q}}{Z} \int_0^{\theta_A} \theta^2 \left(1 + \frac{q\theta^2}{2} \right) I_1(p\theta) d\theta, \qquad (4.9)$$

under the assumption that $|q\theta^2| \ll 1$. The above equation can be further simplified using the relationship among Bessel functions Eq. (4.7). After some calculation, a state equation for smectic-phase monolayer films is obtained as

$$\theta_c = a^{1/2} \sigma \left(\frac{g_y \theta_c \mu^2}{akT} \right) \left\{ 1 - g_z a^{-1/2} \frac{\mu^2}{kT} \left[\frac{1}{2} - \frac{akT}{g_y \theta_c \mu^2} \sigma \left(\frac{g_y \theta_c \mu^2}{akT} \right) \right] \right\} \equiv f(\theta_c), \qquad (4.10)$$

where $\sigma(z) = I_2(z)/I_1(z)$. In order to examine the polar orientational phase transition in smectic monolayers, the function $\sigma(z)$ in Eq. (4.10) is explained as a Taylor series in the vicinity of $z = 0$

$$\sigma(z) = \frac{1}{4}z - \frac{1}{96}z^3 - \frac{5}{1536}z^5 + \cdots \cdots . \qquad (4.11)$$

Substituting the above relation into Eq. (4.10), it is possible for one to obtain the polynomial expansion of $f(\theta_c)$

$$f(\theta_c) = \frac{1}{4} \frac{g_y \mu^2}{a^{1/2}kT} \left[1 - \frac{1}{4} \frac{g_z \mu^2}{a^{1/2}kT} \right] \theta_c - \frac{a^{1/2}}{96} \left(\frac{g_y \mu^2}{akT} \right)^3 \theta_c^3 - \frac{5a^{1/2}}{1536} \left(\frac{g_y \mu^2}{akT} \right)^5 \theta_c^5 + \cdots \cdots . \qquad (4.12)$$

As the second and the third coefficients of the expansion Eq. (4.12) have the same sign, it is well established that the polar orientational transition between a normal-director phase and a tilted-director phase in smectic monolayers cannot be a first-order transition. In other words, a second-order transition is the most reasonable candidate for such a transition. The general criterion for such a second-order transition, induced by monolayer compression, is given by

$$\frac{1}{4} \frac{g_y \mu^2}{a_c^{1/2}kT} \left[1 - \frac{1}{4} \frac{g_z \mu^2}{a_c^{1/2}kT} \right] = 1, \qquad (4.13)$$

that is

$$a_c = \frac{g_y^2 \mu^4}{64k^2T^2} \left(1 + \sqrt{1 - 4g_z/g_y} \right)^2 . \qquad (4.14)$$

This is the critical area for monolayer films where the phase transition between a normal-director phase and a tilted-director phase occurs. The relationship between the molecular configuration of monolayer films and the phase transition is clarified. It is clearly indicated from Eq. (4.12) that the tilted-director smectic phase $(a < a_c)$ is present ahead of the normal-director smectic phase $(a > a_c)$ during the process of monolayer compression. A variety of orientational phase transitions, for example,

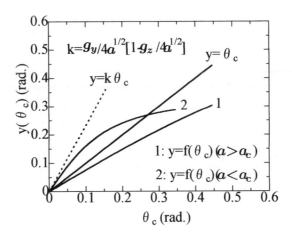

Figure 4.2 Numerical calculation of the self-consistent equation $\theta_c = f(\theta_c)$ of a tilted-director phase for smectic monolayers. $k = 1$ corresponds to the second-order orientational transition between a normal-director phase and a tilted-director phase induced by the monolayer compression.

polar smectic-A–smectic-C, polar smectic-B–smectic-H, etc., can be described by such a relation by only changing the g parameters (g_y and g_z). Figure 4.2 shows the numerical calculation of the present argument. The molecular configuration is assumed to be unchanged in the monolayer compression process, that is, g_y and g_z are independent of the molecular area A. As an example, $g_y = 3$ and $g_z = 0.4$ are set in Fig. 4.2. The values of g_y and g_z are chosen on the basis of the Maxwell-displacement-current (MDC) measurement on 4-cyano-4'-5-alkyl-biphenyl by the authors. [21] It can be seen from Fig. 4.2 that when the slope k of $y = f(\theta_c)$ at the origin becomes 1, a second-order phase orientational transition occurs. If $g_y \gg 4g_z$, one gets a critical area which is simply expressed as $a_c = g_y^2 \mu^2 / 16kT$. From the condition that $a_c < 1$ and a_c is a real quantity, a necessary condition on molecular configuration for possible polar orientational phase transition between a normal-director phase and a tilted-director phase in monolayer films [18]

$$\frac{4kT}{\mu^2} - \frac{16k^2T^2}{g_y\mu^4} < g_z < \frac{g_y}{4}.$$

is obtained. The condition as expressed in the above equation suggests that the molecular configuration having the polar orientational phase transition should be the one capable of producing large enough in-plane electric field in one direction (with large enough g_y). This involves a competition between two electric fields: the electric field normal to the smectic monolayer films (depending on g_z) and the in-plane electric field (depending on g_y). Some kinds of molecular configurations may not exist at the orientational transition between a normal-director phase and a tilted-director phase. The following section will be devoted to the quantitative calculation of the orientational order parameters and the dielectric constant in the normal-director smectic phase and the tilted-director smectic phase respectively.

4.3 Change of Orientational Order Parameter at the Critical Point

To gain more insight into the nature of the polar orientational phase transition in smectic monolayer films, the interaction energy of constituent molecules is used to calculate the orientational order parameters and the dielectric constant of monolayer films in the normal-director smectic phase and the tilted-director smectic phase quantitatively. For simplicity, only the dielectric isotropic case is considered in this section.

4.3.1 Orientational Order Parameter

In order to study the orientational picture of tilted-director smectic monolayers, one needs some order parameters to describe the orientational feature of monolayer films. A reasonable assumption is made that the orientational distribution of the constituent molecules of the monolayer film discussed here is ruled by Boltzmann statistics. The orientational state of the constituent molecules in the monolayer film can be represented by two orientational order parameters $S_1 = \langle \cos \theta \rangle$ and $\langle \cos^2 \theta \rangle$ $[=(2S_2+1)/3]$ which are defined as the thermal averages over $[0, \theta_A]$ (with respect to θ) and $[0, 2\pi]$ (with respect to ϕ)

$$S_1 = \langle \cos \theta \rangle = \frac{1}{Z} \int_0^{\theta_A} \int_0^{2\pi} \cos \theta \exp \left\{ -W_t(\theta, \phi)/kT \right\} d\phi \sin \theta d\theta$$

$$\langle \cos^2 \theta \rangle = \frac{1}{Z} \int_0^{\theta_A} \int_0^{2\pi} \cos^2 \theta \exp \left\{ -W_t(\theta, \phi)/kT \right\} d\phi \sin \theta d\theta, \qquad (4.15)$$

where k is the Boltzmann constant, T is the temperature, and $W_t(\theta, \phi)$ is the interaction energy acting on the molecules, including the intermolecular interaction and the external interaction applied to the molecules, for example, the external electric field. Z is the single-partition function given by Eq.(4.6). For convenience, another order parameter S_2 defined as Eq. (1.1) for monolayer films is also utilized. A special case of the first orientational order parameter S_1 in Eq. (4.15) when there is no dipole-dipole interaction $[W_t(\theta, \phi) = 0]$ is [9]

$$S_0 = \frac{1 + \cos \theta_A}{2}, \qquad (4.16)$$

which is in fact the first orientational order parameter of monolayer films whose constituent molecules are only subjected to hardcore interaction $\{W_t(\theta, \phi) = 0$ and $\theta \in [0, \theta_A]\}$.

4.3.2 Normal-director Smectic Phase $(A/A_0 > a_c)$

In normal-director smectic phase, the constituent molecules produce an internal electric field E_z normal to the monolayer surface, that is, $g_y = 0$. This is because macroscopically normal-director smectic phase has C_∞ symmetry and no electric fields are produced except E_z. Thus the dipole at the origin is supposed to have interaction energy

$$W_n(\theta, \phi) = -\mu E_z \cos \theta.$$

Here n denotes the normal-director smectic phase. From Eq. (2.15), the polarization per molecule

$$m_z = \mu S_1 + \alpha E_z \tag{4.17}$$

is immediately obtained for an isotropic monolayer, considering the molecule-molecule interaction. Here if the electronic polarizability α is given by Gaussian unit, then a factor of $4\pi\epsilon_0$ should be added. The polarization m_z will create a mean field, which is nothing but E_z and depends on molecular configuration:

$$E_z = -g_z a^{-3/2} m_z, \tag{4.18}$$

where g_z has been defined by Eq. (2.21). Combining Eq. (4.17) and Eq. (4.18) and letting the external field $E = 0$ $(S_E = S_1)$, it is easy for one to obtain the relationship between the internal electric field E_z and the orientational order parameter S_1

$$E_z = -\frac{\mu g_z a^{-3/2}}{1 + \alpha g_z a^{-3/2}} S_1, \tag{4.19}$$

where α is the electronic polarizability for dielectric isotropic monolayer films. On the other hand, the integrated result of the first orientational order parameter S_1 by Eq. (4.15) is a Langevin function, which can be approximately expressed as [15]

$$S_1 = \langle \cos \theta \rangle_n = S_0 + \frac{\mu E_z}{12kT}(1 - \cos \theta_A)^2, \tag{4.20}$$

where S_0 is the first orientational order parameter of monolayers without any interactions [Eq. (4.16)], and n stands for the normal-director smectic phase. Combining Eqs. (4.19) and (4.20), the first orientational order parameter S_1 ($\langle \cos \theta \rangle$) is derived

$$\langle \cos \theta \rangle_n = S_0 - \frac{\mu^2 g_z S_0 a^{-3/2}}{12kT(1 + \alpha g_z a^{-3/2})}(1 - \cos \theta_A)^2, \tag{4.21}$$

which establishes under the approximation $\mu E_z / kT$. To get a quantitative sense, the actual calculation of the first orientational order parameter in the normal-director

parameters	normal-director phase						
a	0.40	0.45	0.50	0.55	0.60	0.65	0.70
θ_c	0	0	0	0	0	0	0
θ_A	0.685	0.735	0.785	0.835	0.886	0.938	0.991
$\langle \cos\theta \rangle$	0.884	0.867	0.849	0.830	0.810	0.789	0.767
$\langle \cos^2\theta \rangle$	0.785	0.756	0.727	0.698	0.668	0.639	0.605
$\langle \Delta S_1^2 \rangle (\times 10^{-3})$	4.23	5.55	7.13	9.01	11.2	13.9	17.0
$\delta\epsilon_{ori}(\times 10^{-4})$	0.727	0.944	1.20	1.52	1.90	2.38	3.00
parameter	tilted-director phase						
a	0.10	0.15	0.20	0.25	0.30	0.35	0.40
θ_c	0.240	0.270	0.284	0.286	0.277	0.255	0
θ_A	0.322	0.398	0.464	0.524	0.580	0.633	0.685
$\langle \cos\theta \rangle$	0.972	0.957	0.941	0.926	0.910	0.895	0.884
$\langle \cos^2\theta \rangle$	0.946	0.917	0.887	0.858	0.831	0.804	0.785
$\langle \Delta S_1^2 \rangle (\times 10^{-4})$	2.18	4.95	9.03	14.6	21.7	30.8	42.3
$\delta\epsilon_{ori}(\times 10^{-5})$	0.334	0.848	1.60	2.60	3.84	5.37	7.27

Table 1: Calculated results of tilt angle θ_c, orientational order parameters $\langle \cos\theta \rangle$ and $\langle \cos^2\theta \rangle$, orientational fluctuation $\langle \Delta S_1^2 \rangle = \langle (\cos\theta - \langle \cos\theta \rangle)^2 \rangle$, and the additional dielectric constant $\delta\epsilon_{ori}$ in the normal-director phase and the tilted-director phase respectively.

phase, as shown in Table 1, was performed through a digital repeat operation of the super equation about S_1

$$S_1 = \frac{1}{Z} \int_0^{\theta_A} \int_0^{2\pi} \cos\theta \exp\left\{ -\frac{\mu^2 g_z a^{-3/2} S_1}{(1 + \alpha g_z a^{-3/2})kT} \cos\theta \right\} d\phi \sin\theta d\theta, \qquad (4.22)$$

where Z is given by Eq. (4.6). From Eq. (4.22), the relationship between S_1 and a can be given quantitatively and discretely by assuming certain molecular parameters. Here $\mu = 0.8$ D, $T = 300$ K, $\alpha = 0.8$ Å^3, $g_y = 1.78 \times 10^{39}$ m^{-3}, $g_z = 2.38 \times 10^{38}$ m^{-3}, and $l = 10$ Å are chosen as sample values [15, 18]. The results shown in Table 1.

4.3.3 Tilted-director Smectic Phase ($A/A_0 < a_c$)

In the tilted-director smectic phase, the director of the monolayer tilts away from the normal direction with an angle θ_c, which is determined by the state equation Eq. (4.10). The interaction energy between the dipole moment μ at the origin discussed and the interaction field Eq. (4.1) induced by its neighboring molecules in a 2D lattice (which is expressed by g_y and g_z) is given by

$$W_t(\theta, \phi) = -(g_y \tan\theta_c \sin\theta \cos\phi - g_z \cos\theta) a^{-3/2} \mu m_z$$

$$= -(g_y \tan \theta_c \sin \theta \cos \phi - g_z \cos \theta)a^{-3/2} \frac{\mu^2 S_1}{(1 + \alpha g_z a^{-3/2})}. \quad (4.23)$$

The first orientational order parameter in this phase, similar to that in the normal-director phase, is the solution of the super equation

$$S_1 = \frac{1}{Z} \int_0^{\theta_A} \int_0^{2\pi} \cos \theta \exp \left\{ \frac{(g_y \tan \theta_c \sin \theta \cos \phi - g_z \cos \theta)\mu^2 a^{-3/2} S_1}{(1 + \alpha g_z a^{-3/2})kT} \right\} d\phi \sin \theta d\theta.$$
$$(4.24)$$

The result is also shown in Table. 1 and Fig. 4.3. From Fig. 4.3, it is found that

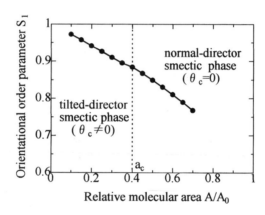

Figure 4.3 Orientational order parameter S_1 of smectic monolayers with respect to the relative molecular area $a = A/A_0$.

the polar orientational transition between the normal-director phase and the tilted-director phase induced by the monolayer compression is a second-order one and it is predicted that the MDC will experience a sudden change at the transition point, as the MDC is proportional to the differentiation of the orientational order parameter S_1 with respect to the molecular area a.

4.4 Dielectric Properties Influenced by the Orientational Phase Transition

According to Eq. (4.12), the polar orientational phase transition induced by mono-layer compression in smectic monolayer films is a second-order one. This is analogous to that in LCs. [5] Nevertheless, such a phase transition in smectic monolayer films is different from the orientational phase transition (e.g. smectic C-smectic A) in LCs in that in monolayer films it depends on in-plane molecular configuration, a characteristic of monolayer films. [17] The dependence of physical quantities such as the average dipole moment in the normal direction (proportional to the first orientational order parameter S_1) or the dielectric constant on the molecular area A, also demonstrates the feature of monolayer films. The influence of the molecular configuration and the molecular area on the dielectric constant comes from the structural difference between monolayers and LCs, i.e., the 2D monolayers and bulk LCs.

4.4.1 Orientational Order Parameter S_1

The first orientational order parameter S_1 of smectic monolayer films in normal-director phase can be viewed as a special case of the tilted-director smectic phase when $\theta_c = 0$. That is, in the case when $\theta_c = 0$, Eq. (4.24) degenerates to Eq. (4.22) and the tilted-director smectic monolayer films return to normal-director smectic monolayer films. Figure 4.3 which is plotted based on Table 1 shows the first orientational order parameter S_1 in the normal-director phase and the tilted-director phase for smectic monolayers. The first orientational order parameter increases gradually in the normal-director smectic phase followed by a sudden decrease of slope, as the monolayer film is compressed to the critical molecular area a_c. The continuity and the abrupt change of the slope of the orientational order parameter S_1 at the critical molecular area a_c in Fig. 4.3 clearly indicates that the monolayer film undergoes a second-order polar orientational phase transition at a_c during the monolayer compression process.

4.4.2 Dielectric Constant

In order to study the polar phase transition, it is also necessary to examine the change in the dielectric constant at the critical point. The additional dielectric constant $\delta\epsilon_{ori}$ due the molecular orientation [the second term in the brace bracket in Eq. (2.19)] is given by

$$\delta\epsilon_{ori} = \frac{\mu^2}{3kTAh\epsilon_0(1 + \alpha_{eff}g_z a^{-3/2})}\langle(\cos\theta - \langle\cos\theta\rangle)^2\rangle. \tag{4.25}$$

Figure 4.4 shows the calculation result of the additional differential dielectric constant $\delta\epsilon_{ori}$ in Eq. (4.25). The simulation result shows that the change of the dielectric constant of monolayer films at the phase transition point is trivial compared with the change of the first orientational order parameter S_1 in Fig. 4.3. A major source for this feature is the dependence of the additional dielectric constant $\delta\epsilon_{ori}$ on the orientational fluctuation $\langle\Delta S_1^2\rangle$, which is a second order compared to the first orientational order parameter S_1. From Fig. 4.4, it is also found that the additional dielectric constant due to orientatioal effect decreases as the monolayer film is compressed, which is reasonable as the possible space for molecular orientation becomes small and the orientational effect disappears gradually as a result of monolayer compression.

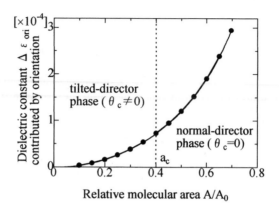

Figure 4.4 Dielectric constant of monolayer films contributed by the molecular polar orientation under monolayer compression. No distinct change is found at the critical point a_c compared with the first orientational order parameter S_1.

For orientational order parameters, an extreme case of zero intermolecular interaction gives Eq. (4.16)

$$\langle \cos^2 \theta \rangle = \frac{1}{3}(1 + \cos \theta_A + \cos \theta_A^2). \tag{4.26}$$

With Eq. (4.16) and Eq. (4.26), it is easy for one to obtain the differential dielectric constant from Eq. 4.25

$$\epsilon_{zz} = 1 + \frac{1}{Ah\epsilon_0(1 + \alpha g_z a^{-3/2})} \left[\frac{\mu^2}{12kT}(1 - \cos \theta_A)^2 + \alpha \right]. \tag{4.27}$$

The property of the orientational fluctuation dependence can be thought to be the major source for the fact that the orientational dielectric constant is described by $(1 - \cos \theta_A)^2$ law, which is also obtained as one result in the calculation of the apparent dielectric constant ϵ_s by the authors [22].

Another major and striking feature of the additional dielectric constant in Eq. (4.25) due to polar orientation is that it can return to the case of bulk state. In the extreme random case, $S_1 = S_2 = 0$, the dielectric constant Eq. (2.19) degenerates to

$$\epsilon_{zz} = 1 + \frac{1}{Ah\epsilon_0(1 + \overline{\alpha} g_z a^{-3/2})} \left[\frac{\mu^2}{3kT} + \left(\overline{\alpha} + \frac{2}{3}\Delta\alpha \right) \right], \tag{4.28}$$

of which the orientational part is proportional to $\mu^2/3kT$. This is nothing but the characteristic of bulk materials. In this case, the dielectric anisotropy brings about an additional electronic polarizability $2\Delta\alpha/3$ to the differential dielectric constant for monolayer films. $\bar{\alpha}$ and $\Delta\alpha$ are the average electronic polarizability and the anisotropy defined by $\bar{\alpha} = (\alpha_{\parallel} + 2\alpha_{\perp})/3$ and $\Delta\alpha = \alpha_{\parallel} - \alpha_{\perp}$, respectively.

For a full understanding of the orientational phase transition process discussed above, it is crucially important to find an experimental way to determine the orientational phase transition point. From the dielectric constant in the normal-director phase and the tilted-director phase (Fig. 4.4), it can be found that the change of the differential dielectric constant is trivial compared with the change of the first orientational order parameter S_1. Observing such a phase transition by dielectric constant therefore can be expected to be difficult. Fortunately, the change of the first orientational order parameter, though being second order, is explicit. This implies that the critical area a_c of the polar orientational phase transition induced by monolayer compression can be measured by, for example, the Maxwell-displacement-current (MDC) measuring technique [23], as the first orientational order parameter S_1 could be connected to the MDC given by Eq. (3.4). [24] From the abrupt change of experimental MDC, it is possible to find out the orientational phase transition point $a = a_c$. If such an argument holds, one might expect that the MDC technique is a good measuring technique to observe the critical molecular area a_c of the polar orientational phase transition, and will support the prediction by the present theory.

The calculation of intermolecular electrostatic interaction in this chapter is a discrete treatment. The operation of the internal electric field at the origin discussed is implemented by summing up the electric field produced by its neighboring molecules in a 2D discrete lattice. This is different from the integral treatment used in the domain calculation by McConnell group. [10] They presented a theory that the shapes of 2D solid domains of phospholipid are determined by a competition between repulsive electrostatic forces and interfacial line tension. The domain shape determination as developed by McConnell *et al.* is in fact a quasi-continuum calculation and the molecular configuration was not considered. Therefore in the future study of configurational phase transition, the problem of describing the molecular configuration remains if one adds the elastic energy which is based on the continuum theory, to the electrostatic energy.

Here it should be pointed out that in the real monolayer films under monolayer compression, a sequence of phase transitions, including the polar orientational phase transition discussed in this chapter and the configurational phase transition, may occur. Understanding the whole process would require an extensive study on the configurational phase transition and other phase transitions occurring in the whole compression process. Nevertheless, the introduction of the self-consistent theory to the polar orientational phase transition for smectic monolayer films is clearly a major step forward for a profound understanding of the smectic monolayer films.

4.5 Summary

In this chapter, the differential dielectric constant was defined and a general expression was given for monolayer films. The orientational effect on the differential dielectric constant is found to be proportional to the orientational fluctuation and decreases steadily as the monolayer films are compressed. The polar orientational phase transition induced by monolayer compression was examined and found to be a second-order one. It is an important implication that only monolayer films with certain kinds of molecular configurations can have such a phase transition from the necessary condition on g_y and g_z. It was also found that the difference of the dielectric constant between normal-director smectic monolayers and tilted-director smectic monolayers in free space is trivial compared with the change of the orientational order parameter S_1. This reveals that the MDC technique is an appropriate candidate for observing such a phase transition.

References

[1] S. Chandrasekhar, *Liquid Crystals*, Cambridge University Press, London (1977).

[2] P. G. de Gennes, *The Physics of Liquid Crystals*, Clarrendon, Oxford (1991).

[3] A. Rapini, *J. Phys. (Paris)*, **33** (1972) 237.

[4] T. Carlson, I. W. Stewart, and F. M. Leslie, *J. Phys. A*, **25** (1992) 2371.

[5] F. Gießelmann and P. Zugenmaier, *Phys. Rev.*, **E55** (1997) 5613.

[6] J. Schacht, Gießelmann, P. Zugenmaier, and W. Kuczyński, *Phys. Rev.*, **E55** (1997) 5633.

[7] O. Albrecht, H. Gruler, and E. Sackmann, *J. Phys. (Paris)*, **39** (1978) 301.

[8] X. Qiu, J. Ruiz-Garcia, K. J. Stine, C. M. Knobler, and J. V. Selinger, *Phys. Rev. Lett.*, **67** (1991) 703.

[9] A. Sugimura, M. Iwamoto, and Z. C. Ou-Yang, *Phys. Rev.*, **E50** (1994) 614.

[10] D. J. Keller, H. M. McConnell, and V. T. Moy, *J. Phys. Chem.*, **90** (1986) 2311.

[11] Z. Cai and S. A. Rice, *Faraday Discuss. Chem. Soc.*, **89** (1990) 211.

[12] Z. Cai and S. A. Rice, *J. Chem. Phys.*, **96** (1992) 6229.

[13] G. M. Sessler, *Electrets*, Springer Verlag, New York (1987).

[14] A. Valance and C. Misbah, *Phys. Rev.*, **E55** (1997) 5564.

[15] C. X. Wu and M. Iwamoto, *Phys. Rev.*, **B55** (1997) 10922.

[16] D. M. Taylor and G. F. Bayes, *Phys. Rev.*, **E49** (1994) 1439.

[17] R. E. Collin, *Field Theory of Guided Waves*, McGraw-Hill, New York (1960) Chap. 12.

[18] C. X. Wu and M. Iwamoto, *Phys. Rev.*, **E57** (1998) 5740.

[19] J. N. Israelachvili, *Intermolecular and Surface Forces*, Academic, London (1985).

[20] I. S. Gradsteyn and I. M. Ryzhik, *Table of Integrals, Series and Products*, Academic, Orlando (1980) formulas 8.431 and 8.486.

[21] M. Iwamoto, T. Kubota, and M. R. Muhamad, *J. Chem. Phys.*, **102** (1995) 9368.

[22] M. Iwamoto, Y. Mizutani, and A. Sugimura, *Phys. Rev.*, **B54** (1996) 8186.

[23] M. Iwamoto, Y. Majima, H. Naruse, T. Noguchi, and H. Fuwa, *Nature*, **353** (1991) 645.

[24] M. Iwamoto and Y. Majima, *J. Chem. Phys.*, **94** (1991) 5135.

CHAPTER 5

DIELECTRIC RELAXATION PHENOMENA

The transient phenomena always attract scientists' attention. Over the past several decades, the dielectric relaxation phenomena in dielectric materials including organic materials, lipids, and liquids have been a subject of many studies in various fields: physics, chemistry, electronics, biology, and the like. One of the most important contributions in these fields was the idea developed by Debye in 1913. He studied the rotational Brownian motion of molecules with permanent electric dipoles in liquids and developed a method for the analysis of the dielectric relaxation phenomena in bulk materials. [1] On the basis of Debye philosophy, many studies of the dielectric relaxation phenomena such as polarization and depolarization in dielectric bulk materials have been carried out for the past decades. [2] Recently these studies have been applied to the investigation of the dielectric properties of ultrathin films such as organic mono- and multilayer films, assuming *a priori* that the dielectric behavior of polar molecules in monolayer and multilayer films is the same as that of polar molecules in isotropic bulk materials. [3] Tanguy and Hesto studied the polarization effects associated with the initial heating of orthphenanthrolin multilayers sandwiched between metal electrodes. [3] The investigation by Jonscher on thermally stimulated depolarization current (TSD) flowing across organic multilayers on the basis of the so-called "Universal law" lead to a different theoretical approach to TSD. [4] Unfortunately, following the Debye model, few theoretical expressions were successfully constructed for analyzing the dielectric relaxation phenomena in monolayers on a material surface. Further it was very difficult to obtain the reproducibility in the TSD measurement: particularly in the measurement of one-layer film sandwiched between top and base electrodes, principally due to the destruction of film by the application of the top electrode.

In the last several chapters, the static properties for monolayers have been discussed. In this chapter, focus will be shifted to the transient phenomena in monolayer films, i.e., the dielectric relaxation phenomena in monolayer films at an air-liquid interface by monolayer compression. That the monolayer structure is changeable under compression and that the molecular motion of rodlike molecules on a material surface is spatially restricted in the region of the semisphere, owing to the presence of the material surface, provide unique and interesting transient phenomena for monolayers. The Maxwell displacement current (MDC) measuring technique [5] is in fact a transient current generated across an electrode suspended in air and a material surface, e.g., due to the orientational change in polar molecules on the material surface. The MDC measurement can give essential information on the polar orientational order S_1 in monolayers at the liquid-air interface even in the range of immeasurably low surface pressure. Thus the technique is suitable for studying the dielectric relaxation phenomena accompanying the orientational ordering and disordering of monolayers on a material surface, where the relaxation phenomena are induced by external stimulation, such as monolayer compression and photoirradiation. Further, a classical mean field approach [6] is used to calculate the Maxwell displacement current due to the ordering and disordering of organic multilayers, under the assumption that the orientational change in the constituent molecules is quick enough to respond to the external stimulation, such as surface pressure application and temperature. This approach is unfortunately restricted within the frame work of equilibrium thermodynamical statistics and it is necessary to combine the Debye method [7] with some monolayer model [6] for better understanding of the dielectric relaxation phenomena in monolayers. For example, nonequilibrium phenomena which happen at the onset of phase transition will be clarified. In this chapter, the transient process of monolayers on a water surface by monolayer compression on the basis of the Debye rotational Brownian motion equation [7] will be discussed with a derivation of the dielectric relaxation time. The orientational order parameter of monolayers during the dielectric relaxation of monolayers is then rectified, with consideration of relaxation time τ and the monolayer viscosity ξ. Further, the generation of Maxwell displacement current (MDC) across 4-cyano-4'-5-alkyl-biphenyl(5CB) Langmuir film in the range of immeasurably low surface pressure at the molecular area close to the onset phase transition from the isotropic planar surface alignment phase to the polar orientational one by monolayer compression will be examined. Finally a method for determining the dielectric relaxation time based on the Debye Brownian motion model will be described.

5.1 Rotational Debye Brownian Motion Model

Figure 5.1 shows a model of a monolayer on a water surface, which is used for the following analysis. The monolayer consists of rodlike polar molecules with a length l. Each molecule has a permanent dipole moment μ in the direction along its long axis, and it stands on a water surface at a tilt angle θ away from the normal direction to the water surface. The orientational distribution of the constituent molecules is ruled by Boltzmann statistics. It is convenient here to start from the rotational Debye-Brownian equation [7]

$$\frac{\partial \omega(\theta, t)}{\partial t} = \frac{1}{\xi \sin \theta} \frac{\partial}{\partial \theta} [\sin \theta (kT \frac{\partial \omega}{\partial \theta} + \frac{\partial W(\theta, t)}{\partial \theta} \omega(\theta, t))], \tag{5.1}$$

where $\omega(\theta, t)$ is the possibility function representing the possibility of the molecules standing on a water surface at a tilt angle θ at time t. ξ is the friction constant of monolayer, and $W(\theta, t)$ is the interaction energy working on molecules. $W(\theta, t)$ contains the dipole-surface interaction $W_s(\theta)$, dipole-dipole interaction $W_d(\theta)$ and the additional interaction $W_{ex}(\theta, t)$ produced by external stimulations such as the electric field. In the present analysis, only dipole-dipole interaction and equivalent additional

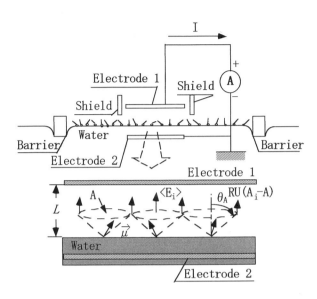

Figure 5.1 Sketch of Debye Brownian motion model of Langmuir film at an air-liquid interface.

piezoelectrical interaction energy are taken into consideration in the mean field approximation, because the dipole-surface interaction is negligibly small in comparison with the other two interactions in the region of the molecular area $A < A_0$. [8] Here A_0 is the critical molecular area $(A_0 = \pi l^2)$ at which molecules lying on a water surface due to the dipole-water surface interaction stand up by monolayer compression. The first term on the right hand side of Eq. (5.1) represents the possibility flow due to the thermal motion of the environment obeying the Fick law. The second term is the flow due to the effect of internal interaction working among molecules and external interactions produced by monolayer compression. Equation (5.1) is nothing but a continuity equation of possibility flux flow, which can be rewritten as

$$\frac{\partial \omega}{\partial t} + \nabla \cdot \mathbf{j} = 0. \tag{5.2}$$

Here $\mathbf{j}$ is the possibility flux flow given by

$$\mathbf{j}(\theta_A, t) = -[\frac{kT}{\xi}\frac{\partial \omega}{\partial \theta} + \frac{\omega}{\xi}\frac{\partial W(\theta, t)}{\partial \theta}]\mathbf{e}_\theta. \tag{5.3}$$

It is understandable that only the portion along direction $\mathbf{e}_\theta$ exists because of the rotational motion of molecules on a material surface, as shown in Fig. 5.1. Under a mean field approximation, it is reasonable to assume that the motion of rodlike polar molecules is restricted within $0 < \theta < \theta_A$, where $\theta_A = \sin^{-1}\sqrt{A/A_0}$, principally due to effects of hardcore intermolecular repulsive forces. [6] Thus at $\theta = \theta_A$, the following boundary condition concerning the possibility flux flow is satisfied:

$$\mathbf{j}\,|_{\theta=\theta_A} = \mathbf{0}. \tag{5.4}$$

At the equilibrium when no external stimulus is applied, there is no possibility flux flow among molecules. Thus the possibility flux $\mathbf{j}(\theta, t)$ should be

$$\mathbf{j}(\theta, t) = -[\frac{kT}{\xi}\frac{\partial\omega}{\partial\theta} + \frac{\omega}{\xi}\frac{\partial W_{\text{int}}(\theta, t)}{\partial\theta}]_{\text{eq}}\mathbf{e}_\theta = 0, \tag{5.5}$$

that is,

$$\frac{kT}{\xi}\frac{d\omega}{d\theta} = -\frac{\omega}{\xi}\frac{\partial W_{\text{int}}(\theta, t)}{\partial\theta}. \tag{5.6}$$

From Eq. (5.6), we obtain

$$\omega(\theta) \sim \exp\{-\frac{W_{\text{int}}}{kT}\}. \tag{5.7}$$

This is the Boltzmann distribution we used for the calculation of the dielectric constant of monolayers, [8] TSD current across monolayers, [6] and MDC current I_{eq} generated by monolayer compression [5] in the last several chapters.

5.1.1 Dielectric Relaxation Equation for Monolayer Films

With the orientational order parameter of monolayers on a water surface, which is defined as [6]

$$S = \int_0^{\theta_A} \cos\theta\omega(\theta, t)\sin\theta d\theta, \tag{5.8}$$

and Eq. (5.1), it is found that S satisfies the following equation: [9]

$$\frac{\partial S}{\partial t} = \int_0^{\theta_A} \frac{1}{\xi}\sin^2\theta\left(kT\frac{\partial\omega(\theta, t)}{\partial\theta} + \frac{\partial W(\theta, t)}{\partial\theta}\omega(\theta, t)\right)d\theta. \tag{5.9}$$

For convenience, in the following in this chapter, S refers to the first orientational order parameter S_1 if without any further explanation. In order to examine the dielectric relaxation phenomena by monolayer compression, it is assumed here that a step additional interaction $RU(A_i - A)$ is produced in monolayers as a result of piezoelectrical effect, starting from an arbitrary equilibrium state at the molecular area $A = A_i$ by monolayer compression. It should be noted that the study of nonequilibrium phenomena in monolayers on a water surface due to the external compression with the aid of movable barriers requires the consideration of surface pressure propagation in the monolayers (see Fig. 5.1). However, this leads to a complicated calculation which is very difficult to solve. Therefore for simplicity it is assumed that the response of the surface pressure propagation is quicker than that of the dielectric relaxation of monolayers. In this case, for all molecules on a water surface, $W(\theta, t)$ is given by

$$W(\theta, t) = W_{int}(\theta) - RU(t)\cos\theta, \tag{5.10}$$

where $-RU(t)\cos\theta$ is the interaction produced by the external stimulation. Here $U(A_i - A)$ is a unit step function defined as

$$U(A_i - A) = \begin{cases} 1 & (A_i > A) \\ 0 & (A_i < A). \end{cases}$$

Substituting Eq. (5.10) into Eq. (5.3), the possibility flux flow is rewritten as

$$\mathbf{j} = -[\frac{kT}{\xi}\frac{\partial\omega}{\partial\theta} + \frac{\omega}{\xi}\frac{\partial W_{int}}{\partial\theta} + \frac{RU(A_i - A)}{\xi}\sin\theta\omega]\mathbf{e}_\theta. \tag{5.11}$$

In the first order approximation, it is assumed that the external interaction R works as a perturbation to the equilibrium state expressed by Eq. (5.5), and influences the orientational distribution of dipoles slightly. Therefore Eq. (5.11) can be approximately written as

$$\begin{aligned}
\mathbf{j} &\approx -\{\frac{kT}{\xi}\frac{\partial\omega}{\partial\theta} + \frac{\omega}{\xi}\frac{\partial W_{int}}{\partial\theta}\}_{eq}\mathbf{e}_\theta - \frac{RU(A_0 - A)}{\xi}\sin\theta\omega\mathbf{e}_\theta \\
&= -\frac{RU(A_i - A)}{\xi}\sin\theta\omega\mathbf{e}_\theta.
\end{aligned} \tag{5.12}$$

The first term of Eq. (5.12) is the possibility flux flow at the equilibrium state, which is zero. Therefore Eq. (5.9) becomes [9]

$$\begin{aligned}
\frac{\partial S}{\partial t} &= \frac{RU(t)}{\xi} \int_0^{\theta_A} \sin^2\theta\omega(\theta, t)\sin\theta d\theta \\
&\approx \frac{RU(t)}{\xi}(1 - \langle\cos^2\theta\rangle),
\end{aligned} \tag{5.13}$$

where $\langle\cos^2\theta\rangle$ is the thermal average due to the interaction W_{int} at $t = 0$. In addition to the term given in Eq. (5.13), the total rate of change of S has to contain a term due to thermal relaxation motion. The thermal motion will tend to restore the orientational distribution of molecules to the equilibrium distribution in the absence of external stimulation. Introducing $1/\tau$ as a proportionality factor, the total change rate of S is defined as

$$\frac{d(S - S_{eq})}{dt} = -\frac{S - S_{eq}}{\tau} + \frac{1 - \langle\cos^2\theta\rangle}{\xi}RU(t) \tag{5.14}$$

where τ is the dielectric relaxation time, and S_{eq} is the order parameter at the equilibrium state before the application of the external stimulation. Here S_{eq} is the orientational order parameter at $t = 0$, whose distribution is ruled by Eq. (5.7). Equation (5.14) describes the transient dynamics of a monolayer, that is, the dielectric relaxation phenomena in monolayers by the application of the external stimulation. The monolayer experiences an additional interaction $-R\cos\theta$ soon after the application of an external stimulus.

5.1.2 Dielectric Relaxation Time

In the equilibrium state at time $t = \infty$, the orientational distribution of polar molecules can be expressed by Boltzmann distribution function again and is given by

$$
\begin{aligned}
\omega(\theta, \infty) &= \frac{\exp\{-W(\theta)/kT\}}{Z}, \\
&= \frac{\exp\{-W_{int}(\theta)/kT\}}{Z}\left\{1 + \frac{R\cos\theta}{kT}\right\},
\end{aligned}
\tag{5.15}
$$

under the assumption $|R/kT| \ll 1$. Here the partition function Z is expressed as

$$
Z = Z_0 + \frac{R}{kT}S_{eq}Z_0,
\tag{5.16}
$$

where Z_0 is the single-partition function at $t = 0$. From Eqs. (5.15) and (5.16), we obtain the orientational order parameter S at time $t = \infty$ as follows:

$$
\begin{aligned}
S &= \int_0^{\theta_A} \cos\theta\,\omega(\theta, \infty)\sin\theta\,d\theta \\
&= S_{eq} - \frac{R}{kT}(\langle\cos\theta\rangle^2 - \langle\cos^2\theta\rangle),
\end{aligned}
\tag{5.17}
$$

where $\langle\cos\theta\rangle$ and $\langle\cos^2\theta\rangle$ are the thermal averages before external stimulation. At the equilibrium state at $t = \infty$, the relation $d(S - S_{eq}^e)/dt = 0$ should be satisfied. Therefore, from Eqs. (5.14) and (5.17), the dielectric relaxation time is obtained as:

$$
\tau = \frac{\xi}{kT}\frac{\langle(\cos\theta - \langle\cos\theta\rangle)^2\rangle}{1 - \langle\cos^2\theta\rangle}.
\tag{5.18}
$$

Equation (5.18) gives the dielectric relaxation time τ in a generalized form, which is proportional to the orientational fluctuation of monolayers ($\langle(\cos\theta - \langle\cos\theta\rangle)^2\rangle$), and is inversely proportional to the diffusion coefficient ($=kT/\xi$). [7] The denominator ($1 - \langle\cos^2\theta\rangle$) is related to direction of the external stimulation [see Eq. (5.14)]. A special case when the monolayer film is only subjected to hardcore repulsive interaction gives

$$
\tau_0 = \frac{\xi}{4kT}\frac{1 - \cos\theta_A}{2 + \cos\theta_A},
\tag{5.19}
$$

using Eq. (4.16) and Eq. (4.26).

5.1.3 Equivalent Circuit for Transient Relaxation Process

From Eq. (5.14), it is clear that the transient relaxation process by monolayer compression can be expressed using an equivalent circuit model with a circuit response time $\tau(= RC)$, as illustrated in Fig. 5.2. Here the voltage source V, the capacitance

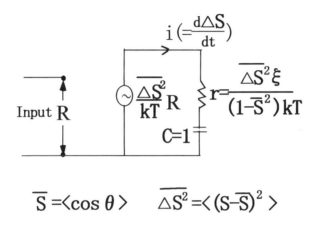

$$\overline{S} = \langle \cos \theta \rangle \qquad \overline{\Delta S^2} = \langle (S - \overline{S})^2 \rangle$$

Figure 5.2 Equivalent circuit model showing the transient relaxation process in monolayer films at the air-liquid interface under monolayer compression.

C, and the time-variable electrical resistance r are defined as represented in the figure. The current I following through the circuit is defined as

$$I \propto \frac{d}{dt}S - \frac{d}{dt}S_{\text{eq}}. \tag{5.20}$$

That is, the current I is just proportional to the difference between the experimental MDC across monolayers, [6] and the MDC proportional to dS_{eq}/dt, which can be calculated under the assumption that the orientational distribution of polar molecules is given by Eq. (4.15). [6] The voltage generated across the capacitance $(C = 1)$ gives the deviation of the experimental orientational order parameter S from the theoretical orientational order parameter S_{eq} at the equilibrium state. As the electrical circuit illustrated in Fig. 5.2 responses to the input-controlled voltage V with a response time $\tau(= RC)$, that is, the monolayer responses to the monolayer compression with a time delay τ, we expect that $\Delta S(= S - S_{\text{eq}})$ should be generated during monolayer compression in the MDC measurement.

5.2 Relaxation Process at an Air-water Interface

5.2.1 Mean-field Approximation

With Eqs. (4.26) and (5.19), Eq. (5.14) is rewritten as

$$\frac{d(S - S_{eq})}{dt} = -\frac{S - S_{eq}}{\tau} + \frac{(2 + \cos\theta_A)(1 - \cos\theta_A)}{3\xi} RU(A_i - A), \qquad (5.21)$$

where τ is the relaxation time. This is the equation describing the transient behavior of the monolayer with small molecular interactions, that is, the dielectric relaxation phenomena in monolayers by monolayer compression. The dielectric relaxation phenomena of monolayers in the mean field approximation will now be explained using Eq. (5.21). However, before going to the discussion, the parameters S_{eq} and R in Eq. (5.21) will be briefly defined.

The monolayer mainly experiences a molecule-molecule interaction W_e when the molecular area A is smaller than A_0, due to monolayer compression. The external stimulation to a monolayer is expressed by introducing a step interaction R starting from the molecular area $A = A_i$ close to A_0 ($A_i < A_0$). The orientational order parameter S in the equilibrium state ($=S_{eq}^e$) at the molecular area A is expressible in a mean field approximation [6]

$$S_{eq}^e = \frac{e^t - \cos\theta_A e^{t\cos\theta_A}}{e^t - e^{t\cos\theta_A}} - \frac{1}{t}, \qquad (5.22)$$

where

$$t = \frac{-\mu\langle E_i \rangle}{kT}.$$

Here, $\langle E_i \rangle$ is the average internal electric field acting on the constituent rodlike polar molecules. A further calculation of Eq. (5.22) leads to the final state satisfying the following relation:

$$S^F - S_{eq}^e = \frac{R}{12kT}(1 - \cos\theta_A)^2, \qquad (5.23)$$

when $| t | << 1$ and $| R/kT | << 1$. At the equilibrium state, the relation $d(S - S_{eq}^e)/dt = 0$ is satisfied. Therefore from Eqs. (5.21) and (5.23),

$$\frac{R}{12kT}(1 - \cos\theta_A)^2 = \frac{\tau(2 + \cos\theta_A)(1 - \cos\theta_A)}{3\xi} R, \qquad (5.24)$$

is obtained. Hence relaxation time τ becomes

$$\tau^e = \frac{\xi}{4kT}\frac{1 - \cos\theta_A}{2 + \cos\theta_A}, \qquad (5.25)$$

which is inversely-proportional to the diffusion coefficient defined as [7]

$$D = \frac{kT}{\xi} \qquad (5.26)$$

Eq. (5.25) is nothing but Eq. (5.19). This means that monolayers subjected to a molecule-molecule interaction have the same dielectric relaxation time as that with hardcore interaction. It is interesting to note here that the relaxation time τ depends on $\cos\theta_A = \sqrt{1 - A/A_0}$, thus the relaxation time τ decreases as the molecular area A decreases by monolayer compression. Substituting Eqs. (5.25) and (5.26) into Eq. (5.21), a differential equation expressing the deviation of the orientational order parameter from the equilibrium state is obtained

$$\frac{d\triangle S}{dt} = -\frac{4kT}{\xi}\frac{2 + \cos\theta_A}{1 - \cos\theta_A}\triangle S + \frac{(2 + \cos\theta_A)(1 - \cos\theta_A)}{3\xi}RU(A_i - A), \qquad (5.27)$$

where $\triangle S = S - S_{eq}^e$. With the relation $\sin\theta_A = \sqrt{A/A_0}$, it is derived that

$$\frac{d(S - S_{eq}^e)}{dA}\frac{dA}{dt} = -\frac{4kT(2 + \sqrt{1 - [A/A_0]})}{\xi(1 - \sqrt{1 - [A/A_0]})}(S - S_{eq}^e)$$

$$+ \frac{(2 + \sqrt{1 - [A/A_0]})(1 - \sqrt{1 - [A/A_0]})}{3\xi}RU(A_i - A). \,(5.28)$$

In order to analyze the dielectric relaxation phenomena starting from the equilibrium state at the molecular area $A = A_i$ close to A_0 ($> A_i$) by monolayer compression, an approximate of Eq. (5.28) may be written as

$$\alpha\frac{d(S - S_{eq}^e)}{dA} = \frac{4kT}{\xi}(2 + 3\sqrt{1 - [A/A_0]})(S - S_{eq}^e)$$

$$-\frac{RU(A_i - A)}{3\xi}(2 - \sqrt{1 - [A/A_0]}), \qquad (5.29)$$

with $dA/dt = -\alpha$, where α in this chapter is the molecular compression speed. When there is no external stimulation, i.e., $R = 0$, $\alpha = 0$, Eq. (5.29) returns to the equilibrium equation $S - S_{eq}^e = 0$. Here it should be noted that the external interaction R is produced as a result of the constant molecule compression at a speed of α. Integrating the differential equation above from the equilibrium state at $A = A_i$ to a nonequilibrium state at $A = A$ with the initial condition

$$S\,|_{A=A_i} = S_{eq}^e\,|_{A=A_i}, \qquad (5.30)$$

the deviation of order parameter from equilibrium state is found to be

$$S - S_{eq}^e\,|_{A=A} = -\frac{2RA_i}{3\xi\alpha}(1 - \frac{A}{A_i})(1 - \frac{A_i - A}{\tau_i}\frac{1}{\alpha}) - (S_{eq}^e\,|_{A=A} - S_{eq}^e\,|_{A=A_i}). \qquad (5.31)$$

Here τ_i is the value of τ at the molecular area $A = A_i$

$$\tau_i \approx \tau_c = \frac{\xi}{8kT}\,.$$

It is obvious from Eq. (5.31) that the deviation from equilibrium state comes from a step external compression stimulation and that the thermal diffusion effect which is expressed as $(A_i - A)/\tau_i$ tends to bring molecules back to a new equilibrium state $S_{eq}^e \mid_{A=A}$.

Figure 5.3 shows an example of the orientational order parameter S when an external step stimulation is applied at the molecular area $A = A_i$, calculated from Eq. (5.31) for $A_i/\tau_i\alpha - 3$, $RA_i/\xi\alpha = 3$, and $A_i/A_0 = 0.95$. As will be shown later in

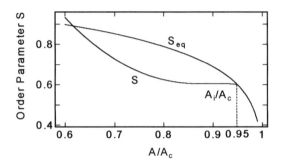

Figure 5.3 Nonequilibrium deviation of order parameter S under an external step stimulation when $A_i/\tau_i\alpha = 3$, $RA_i/\xi\alpha = 3$, and $A_i/A_0 = 0.95$.

Fig. 5.4(b), the experimental results indicate that $S - S_{eq} > 0$ is satisfied in the region of molecule area 65 Å^2 < A < 95 Å^2 = A_0, whereas $S - S_{eq}^e < 0$ is satisfied in the region $A < 65$ Å^2. In other words, the relation $S - S_{eq}^e = 0$ is satisfied at the molecular area close to $A = A_0$ and $A = 2A_0/3$. Therefore, in the calculation, looking at the first term of Eq. (5.31), the above parameters are chosen for the sample calculation. As can be seen in Fig. 5.3, at the initial state, the molecule compression velocity α is greater than $(A_i - A)/\tau$, thus the relation $S - S_{eq}^e < 0$ is satisfied. This means that the response of molecules is relatively slow. While the monolayer is further compressed, the molecular compression speed α becomes smaller than $(A_i - A)/\tau$ because the relaxation time decreases as the monolayer is compressed, as aforementioned. Thus the returning speed to the equilibrium state is so quick that molecules cannot stop at the equilibrium state, which leads to $S - S_{eq}^e > 0$. From Eq. (5.31), the deviation of Maxwell displacement current ΔI from the current I_{eq} which flows so as to satisfy the thermodynamic equilibrium condition may be expressed as

$$\Delta I = I - I_{eq}$$

$$= -\frac{16\mu R}{3\xi L A^2}(1 - \frac{A_i - A}{\tau_c}\frac{1}{\alpha}\frac{A_i + A}{A_i}) \tag{5.32}$$

by substituting Eq. (5.31) into the following equations:

$$I = \frac{B\alpha\mu}{L}(\frac{S}{A^2} - \frac{dS}{AdA}),$$

and

$$I_{eq} = \frac{B\alpha\mu}{L}(\frac{S^{e}_{eq}}{A^2} - \frac{dS^{e}_{eq}}{AdA}),$$

which have been given by Eq. (3.4). Here B is the working area of electrodes and L the distance between Electrode 1 and the water surface (see Fig. 5.1), as defined before.

5.2.2 Closed Packing and Molecule-surface Interaction

In order to calculate the dielectric relaxation time τ given by Eq. (5.18), $\langle\cos\theta\rangle$ (which is equal to S_{eq}) and $\langle\cos^2\theta\rangle$ must be calculated in advance. The mean-field approximation is applicable to the case of monolayers with a small molecular area A ($A < A_0$). In this region, the monolayer is governed by the molecule-molecule interaction W_d, which is written as [8]

$$W_d = \frac{11.0342\mu m_z}{4\pi\epsilon_0 a^3}\frac{2\epsilon_m}{\epsilon_m + 1}\cos\theta, \tag{5.33}$$

under the assumption of the uniform and hexagonal molecular packing. It is the configuration of nearest-neighbor separation distance and hence of the minimum electrostatic interaction energy for any given packing. Here ϵ_m is the relative dielectric constant of monolayer supporting materials, a is the nearest molecular separation distance between adjacent two molecules, and m_z is the average polarized dipole moment of monolayers given by

$$m_z = \frac{\mu S_0}{1 + 11.0342a^{-3}\alpha_e \cdot 2\epsilon_m/(\epsilon_m + 1)}. \tag{5.34}$$

Here α_e is the electronic polarizability of molecules. S_0 is the orientational order parameter $[= (1 + \cos\theta_A)/2]$ in the zero interaction. With Eq. (4.15), the orietational order parameter S^{d}_{eq} ($= \langle\cos\theta\rangle$) and $\langle\cos^2\theta\rangle$ under these interactions are calculated as

$$\langle\cos\theta\rangle = S_0 + \frac{\eta_d}{12}(1 - \cos\theta_A)^2$$

$$\langle\cos^2\theta\rangle = \frac{1}{3}(1 + \cos\theta_A + \cos^2\theta_A) + \frac{\eta_d S_0}{6}(1 - \cos\theta_A)^2, \tag{5.35}$$

where η_d is expressed as

$$\eta_d = -\frac{11.0342\mu m_z}{4\pi\epsilon_0 kT a^3} \cdot \frac{2\epsilon_m}{\epsilon_m + 1}.$$

Here ϵ_0 is the dielectric constant of a vacuum. Substituting Eq. (5.35) into Eq. (5.18),

$$\tau^d = g^d \tau_0 = [1 + \frac{\eta_d S_0 (1 - \cos\theta_A)}{2(2 + \cos\theta_A)}] \cdot \frac{\xi}{4kT} \frac{1 - \cos\theta_A}{2 + \cos\theta_A}. \tag{5.36}$$

Here g^d is given by

$$g^d(\theta_A) = 1 + \frac{\eta_d S_0 (1 - \cos\theta_A)}{2(2 + \cos\theta_A)}, \tag{5.37}$$

an interaction coefficient representing the effect of molecule-molecule interaction, and τ_0 [Eq. (5.19)] is the relaxation time obtained under the condition that both molecule-molecule interaction and molecule-surface interaction are ignored. [9]

In the case of monolayers with the molecular area A close to the critical area A_0, the interaction between molecules and the material surface become very important in comparison with the molecule-molecule interaction. This interaction is given by [11]

$$W_s(\theta) = -\frac{\mu^2}{16\pi\epsilon_0\epsilon_m l^3 \cos\theta} \frac{\epsilon_m - 1}{\epsilon_m + 1}. \tag{5.38}$$

The dielectric relaxation time is mainly affected owing to the presence of the interface. The orientational order parameter S_{eq} $(= \langle\cos\theta\rangle)$ and $\langle\cos^2\theta\rangle$ can also be calculated by substituting Eq. (5.38) into Eq. (4.15), [6]

$$\langle\cos\theta\rangle = \frac{x}{2} + \frac{e^x - \cos^2\theta_A e^{x/\cos\theta_A}}{2[e^x - \cos\theta_A e^{x/\cos\theta_A} + x\{\mathrm{Ei}(x/\cos\theta_A) - \mathrm{Ei}(x)\}]}$$

$$\langle\cos^2\theta\rangle = \frac{1}{3}\{\langle\cos\theta\rangle x + \frac{(2\langle\cos\theta\rangle - x)(e^x - \cos^2\theta_A e^{x/\cos\theta_A})}{e^x - \cos^2\theta_A e^{x/\cos\theta_A}}\}, \tag{5.39}$$

where

$$x = \frac{\mu^2}{16\pi\epsilon_0\epsilon_m l^3 kT} \frac{\epsilon_m - 1}{\epsilon_m + 1}.$$

The coefficient g in this case becomes (see Eq. (5.18) and Eq. (5.19))

$$g^s(\theta_A) = \frac{4(2 + \cos\theta_A)}{1 - \cos\theta_A} \frac{\langle\cos^2\theta\rangle - \langle\cos\theta\rangle^2}{1 - \langle\cos^2\theta\rangle}, \tag{5.40}$$

which can be written as

$$g^s(\theta_A) = \frac{P(\theta_A)}{Q(\theta_A)}, \tag{5.41}$$

with the approximation $|x| \ll 1$. Here

$$\begin{cases} P(\theta_A) = 1 + x\{-\frac{6(1+\cos\theta_A)}{(1-\cos\theta_A)^2} - \frac{4(1+\cos\theta_A+\cos^2\theta_A)}{(1-\cos\theta_A)^2}\ln\cos\theta_A\} \\ \quad\quad + x^2\{-\frac{12}{(1-\cos\theta_A)^2} - \frac{6(1+\cos\theta_A)}{(1-\cos\theta_A)^3}\ln\cos\theta_A\} \\ Q(\theta_A) = 1 + x\{-\frac{3}{2}\frac{1-\cos^2\theta_A}{(2+\cos\theta_A)(1-\cos\theta_A)^2} + \frac{\cos^2\theta_A+\cos\theta_A-5}{(2+\cos\theta_A)(1-\cos\theta_A)^2}\ln\cos\theta_A\} \\ \quad\quad + x^2\{-\frac{12}{(1-\cos\theta_A)^2} + \frac{6(1+\cos\theta_A)}{(1-\cos\theta_A)^3}\ln\cos\theta_A\}. \end{cases}$$

5.2.3 Relaxation Experiment of Monolayer Compression

Based on the above discussion, the change of the orientational order parameter S with respect to molecular area can be investigated using the Maxwell displacement current(MDC) across 5CB monolayers on a water surface at the critical molecular area A_0 by monolayer compression. The experiment was performed using the experimental system shown in Fig. 5.1. [10] The trough has a rectangular shape with an area of 1,050 cm^2. At the center of the trough, electrode 1 with a working area B of 44.2 cm^2 was placed parallel to and at a distance L of 1.0 ± 0.05 mm above the water surface. The two electrodes were short-circuited with a picoampere electrometer. The generation of MDC was monitored together with the surface pressure-area ($\Pi - A$) isotherm during monolayer compression. The experiment was carried out with a molecular compression speed α of 0.054 Å^{2}sec^{-1}/molecule, with the help of two moving barriers (see Fig. 5.1) by moving them at a velocity of 30 mm/min. [10] Figure 5.4(a) gives the results obtained in the experiment and plotted as curve 1 in the figure. From bottom to top, $\Pi - A$ isotherm and MDC-area ($I - A$) isotherm are shown. In the molecular area $A >70$ Å^2, although the surface pressure is very small, the MDC is clearly generated in a manner as previously described by the authors. [10] In the figure, $I - A$ isotherm is plotted as curve 2 when no monolayer is formed on a water surface. In this case surface pressure was nearly zero when two barriers were moved (not shown here). As can be seen in the figure, MDC is initiated to flow at the critical area A_0 of 130 Å^2 by monolayer compression. The orientational order parameter S, which is proportional to the mean vertical component of dipole moment of molecules $m_z = \mu S$, was calculated from the amount of charge flowing across monolayers, [5] and the results were plotted in Fig. 5.4(b). Briefly the results were obtained by integrating the MDC in the range between A_0 to A, and then divided it by A, that is

$$m_z = \int_{A_0}^{A} I dA / A\alpha.$$

In Fig. 5.4(b) the orientational order parameter S_{eq}^e in the equilibrium state is also plotted as a dashed line. As can be found in the figure, the experimental results satisfy the relation $\Delta S < 0$ at the molecular area A close to A_0 ($=130$ Å^2). These results indicate that the barrier velocity $\alpha = -dA/dt$ is large comparing to the response effect $(A_i - A)/\tau_c$ at the transition.

From the Debye-Brownian equation, a nonequilibrium dynamic equation of Eq. (5.21) has been derived for the analysis of the dielectric relaxation phenomena of monolayers on a water surface. This equation is expected to be suitable for the analysis of monolayers on a water surface so long as the rodlike polar molecule model is applicable Expression of the relaxation time given by Eq. (5.25) was also obtained, and we found that the dielectric relaxation depends on the molecular area A. The order parameter under the mean field approximation was given by Eq. (5.31). Any monolayer under an external stimulation will exhibit transient behavior, i.e., experience nonequilibrium intermediate states and finally return to a new equilibrium state. The effect of going

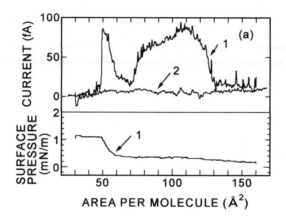

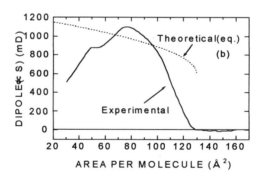

Figure 5.4 (a)Experimental Maxwell displacement current of 5CB with a compression speed of $\alpha = 0.054$ Å^2/sec. Surface pressure-area isotherm (from bottom to top). With monolayers on a water surface (curve 1), and without monolayers on a water surface (curve 2). (b)Experimental mean vertical component of dipole moment of 5CB and theoretical mean vertical component of dipole moment at equilibrium.

back to the new equilibrium state is due to the intermolecular diffusion expressed as kT/ξ in Eq. (5.26). This dielectric relaxation phenomenon yields an additional orientational order parameter $\triangle S$ written by Eq. (5.31) to the equilibrium one S_{eq}. It is also reasonable that the relaxation time τ is inversely proportional to the diffusion coefficient D and depends on the maximum angle θ_A of molecules. Shortly after the onset of transition during monolayer compression, the deviation of order parameter S from S_{eq}^{e} at the equilibrium state is decided by the balance between the compression speed α and the relaxation time τ at the molecule area A close to A_0. As shown in Fig. 5.4(b), at the beginning of the transition, the compression speed is relatively large comparing to the response effect $[\alpha > (A_i - A)/\tau_c]$. Thus the molecules retard to react to such a rapid external change and results in $\triangle S < 0$. After the onset of the transition, the response effect suppresses the compression effect, that is, $\alpha < (A_i - A)/\tau_c$. Thus returning to the equilibrium cannot completely stop due to the inertia effect. This leads to $\triangle S > 0$ at some stages after the onset of transition. In Fig. 5.4(b), $S > S_{\text{eq}}^{\text{e}}$ is satisfied in the region 65 Å^2 < A < 95 Å^2.

The present calculation also reveals the whole relaxation process under the external compression stimulation starting from the molecule area A_i near the critical molecular area A_0. As the monolayer is compressed by an external step stimulation from A_i, the deviation of order parameter S from the order parameter S_{eq}^{e} at the equilibrium state becomes larger, which means that the nonequilibrium effect cannot be negligible. The returning of nonequilibrium state to equilibrium state happens after the deviation of order parameter reaches the maximum. The return speed depends on the balance between speed of compression α and the relaxation speed which is related to the friction constant ξ of monolayer. For further understanding of the dielectric relaxation phenomena, one needs to clarify the piezoelectrical effect represented by R.

5.2.4 Interaction Coefficient

Figure 5.5 shows the interaction coefficient g in the molecular interaction region [g^d in Fig. 5.5(a)] and in the region dominated by the molecule-surface interaction [g^s in Fig. 5.5(b)]. In Fig. 5.5(a), $l = 2$ nm, $T = 300$ K, $\mu = 0.7$ D, and $\alpha_e = 0.65$ Å^3 were chosen as examples. Figure 5.5 was plotted in two extreme cases, that is, $W(\theta) \approx W_{\text{d}}(\theta)$ and $W(\theta) \approx W_{\text{s}}(\theta)$. It was seen that g is less than 1 in both regions, which reveals that the dielectric relaxation time τ becomes shorter due to the molecule-molecule interaction and molecule-surface interaction. The interaction working on molecules, whether it is a repulsive molecule-molecule one or an attractive molecule-surface one, brings a monolayer to a more stable system promptly. As a result the dielectric relaxation time τ decreases. The dielectric relaxation time increases as the molecular area increases in Fig. 5.5(a). This is because in the small molecular area region, the molecule-molecule interaction decreases as the molecular area increases. It is expected that at the molecular area $A_i = A_0 \sin \theta_i$, where the repulsive interactive force changes into an attractive one, $g = 1$, because $W(\theta_i) = W_{\text{d}}(\theta_i) + W_{\text{s}}(\theta_i) = 0$

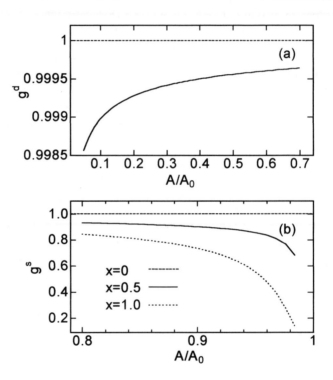

Figure 5.5 Interaction coefficient g with respect to the relative molecular area A/A_0. (a) g^d in the region with a small molecular area, and (b) g^s in the region with a molecular area close to A_0.

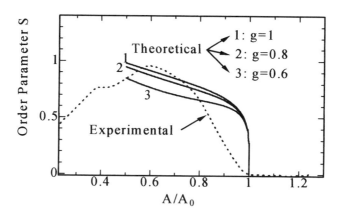

Figure 5.6 Experimental order parameter $S = S_1$ and the theoretical g dependence of the order parameter.

at the molecular area $A = A_i$. In contrast, in the region $A > A_i$, as the molecular area increases, the dielectric relaxation time τ decreases because of the increase in the attractive molecule-surface interaction.

From the definition of g, the dielectric relaxation may also be written in another form

$$\tau = \frac{\xi'}{4kT} \frac{1 - \cos\theta_A}{2 + \cos\theta_A} \qquad (5.42)$$

with an apparent viscosity constant $\xi' = g\xi$. As $g < 1$, the monolayer becomes less viscous as a result of the molecule-molecule interaction and the effect of surface.

Based on the aforementioned discussion, the result of the MDC flowing across 5CB monolayers with monolayer compression in Fig. 5.6 [see Fig. 5.4(a)] is examined. In Fig. 5.6, the results were again plotted using the order parameter S. The interaction coefficient g depends on the molecular area A as shown in Fig. 5.5. However, for simplicity, here the order parameter S using Eqs. (5.27) and (5.31) was calculated, replacing $\xi \to g\xi$ under the assumption that g is constant. This simplification is appropriate, because the change in g is gentle with respect to A (see Fig. 5.5). As seen in Fig. 5.6, it is found that the order parameter saturates more slowly as the interaction increases, i.e., as g deviates from 1. This tendency is favorable for explaining the experimental dielectric relaxation phenomena, though there is still a discrepancy between the theoretical curve and the experimental result at the molecular area close to $A = A_0$, which may be due to the ignorance of the size effect of the constituent molecules at the molecular area close to A_0 [12] and due to the ignorance of the dependence of g on A. The modification of the present model is required. Nevertheless, the generalized relaxation time given by Eq. (5.18) is still correct, and it will be helpful for a profound understanding of the dielectric relaxation phenomena in monolayers on a material surface, as well as the frictional property of monolayers on a material surface. [2, 13]

5.3 Determination of Dielectric Relaxation Time

To gain more insight into the dielectric relaxation phenomena of monolayer films, how to determine the dielectric relaxation time τ for monolayers becomes crucially important. A whole-curve method using MDC, [14] following the similar method developed by Bucci and Fieschi in the thermally stimulated measurement, [15] has been presented for determining the dielectric relaxation time for floating monolayers on the water surface. Such a relaxation time can be expressed as [14]

$$\tau = \frac{\frac{1}{\alpha}\int_{A_i}^{A} I\,dA + \frac{KS_f}{\alpha A}}{I + \frac{1}{A}\int_{A_i}^{A} I\,dA}. \tag{5.43}$$

The first term of the numerator of Eq. (5.43) represents the charge (< 0) flowing through the circuit in the range of molecular areas between A_i and A by monolayer compression. The second term of the numerator is due to the induced charge (> 0) on electrode 1 at the equilibrium state for molecular area A. The first term in the denominator represents the MDC at the molecular area A, and the second term in the denominator is contributed by the average of the MDC in the range of molecular areas between A_i and A. S_f represents the equilibrium order parameter at the molecular area A.

Figure 5.7 gives the relationship between τ and A on the basis of Eq. (5.43) and MDC measurement. It is found that τ increases smoothly as A increases, although a fluctuation in τ is observed, possibly due to the limitations of the experimental measuring system at a molecular area around 110 Å^2, which corresponds to the onset of the phase transition from a planar alignment phase to a polar orientational phase. It should be noted here that the fluctuation is theoretically expected to occur due to the onset of this phase transition. [16] It is also found that the dielectric relaxation time is on the order of 1 to 4000 s, depending on the molecular area A. In the calculation, a columnar model, assuming a rodlike molecule as a column of length l and radius r, was used to replace a rodlike molecule model. [17] As plotted by dotted line, the calculated relaxation time τ agrees well with the experimental relaxation time τ with $l = 0.592$ nm and $r = 0.376$ nm, when the frictional constant $\xi = 5.3 \times 10^{-15}$ Js.

The method employed here to determine the dielectric relaxation time τ of monolayers corresponds to the whole-curve method in the TSC analysis within the framework on dielectric materials by Debye, which has been used to determine the dielectric

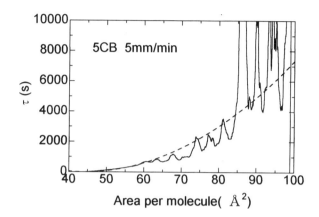

Figure 5.7 Relationship between the relaxation time of the monolayer τ and the molecular area A. The dotted line represents the theoretical relaxation time obtained from Eq. (5.19) with $\xi = 5.3 \times 10^{-15}$ Js.

relaxation time of dipoles in organic films. [18] In the TSC measurement, the short-circuited current flows during heating, due to dipolar depolarization. That is, dipoles are thermally activated as the temperature increases. In other words, dipoles in organic films have more freedom as the temperature increases. On the other hand, in the MDC measurement, the short-circuited current flows by monolayer compression, due to the orientational ordering of polar molecules. The motion of polar molecules is gradually restricted as the monolayer is compressed, because the space available for the rotational motion of molecules decreases as the molecular area decreases. From these discussions, it is found that the temperature axis T in the TSC measurement corresponds precisely to the axis of the molecular area A in the MDC measurement. The basic equation representing the dielectric relaxation process of monolayers in the MDC measurement is similar to that representing the dielectric relaxation process of organic materials in the TSC measurement. Therefore, it may be concluded that the analysis of TSC [2, 19] can be applied to the analysis of MDC after some modifications. For example, as presented here, the concept of the whole-curve method in the TSC measurement for determining the dielectric relaxation time τ of dipoles in dielectric materials can be used for the determination of the relaxation time τ of monolayers. It is well known that TSC results in one peak at a temperature T_m due to dipolar depolarization during heating. Similarly, it is expected that MDC shows one MDC peak at a molecular area A_m during monolayer compression, as shown in Fig. 5.4(a). [20]

5.4 Summary

The dielectric relaxation phenomena of organic monolayers on a water surface under the external compression were investigated on the basis of rotational Debye-Brownian equation. A generalized dielectric relaxation time for monolayer films is derived. The effect of relaxation, which is obtained as Eqs. (5.25) and (5.27) generates a deviation of order parameter $\triangle S$ from that of equilibrium state. The nonequilibrium dielectric relaxation process depends on the balance between the barrier velocity α and the response speed defined as $(A_i - A)/\tau_c$. It was found that the relation $\triangle S < 0$ is satisfied at the onset of transition in the case of the Maxwell displacement current measurement of 5CB monolayers performed in the present experiment. Further the method for determining the dielectric relaxation time was presented based on the similarity between TSC and MDC measurements.

References

[1] P. Debye, *Polar Molecules*, Dover, New York (1929).

[2] H. Fröhlich, *Theory of Dielectrics*, Oxford University Press, New York (1958); R. Chen and Y. Kirsh, *Analysis to Thermally Stimulated Process*, Pergamon, Oxford (1981); J. van Turnhout, *Thermally Stimulated Discharge of Polymer Electrets*, Elsevier Scientific, New York (1975); G. M. Sessler, *Electrets*, Springer-Verlag, Berlin Heidelberg (1987).

[3] V. K. Agarwal, *Electrical Behaviour of Langmuir Films (Electrocomponent Science and Technology)*, Gordon and Breach, New York (1975); J. Tanguy and P. Hesto, *Thin Solid Films*, **21** (1974) 129.

[4] A. K. Jonscher, *J. Phys. D*, **24** (1991) 1633.

[5] M. Iwamoto and Y. Majima, *J. Chem. Phys.*, **94** (1991) 5135; M. Iwamoto, Y. Majima, H. Naruse, T. Noguchi, and H. Fuwa, *Nature*, **353** (1991) 645.

[6] A. Sugimura, M. Iwamoto, and Z. C. Ou-Yang, *Phys. Rev.*, **E50** (1994) 614; M. Iwamoto, C. X. Wu, and W. Y. Kim, *Phys. Rev.*, **B54** (1996) 8191.

[7] J. Mcconnell, *Rotational Brownian Motion and Dielectric Theory*, Academic, New York (1980).

[8] M. Iwamoto, M. Mizutani, and A. Sugimura, *Phys. Rev.*, **B54** (1996) 8186.

[9] M. Iwamoto and C. X. Wu, *Phys. Rev.*, **E54** (1996) 6603.

[10] M. Iwamoto, T. Kubota, and M. R. Muhamad, *J. Chem. Phys.*, **102** (1995) 9368.

[11] C. Kittel, *Introduction to Solid State Physics*, Wiley, New York (1976).

[12] M. Jiang, F. Zhong, D. Y. Xing, Z. D. Wang, and J. Dong, *J. Chem. Phys.*, **106** (1997) 6171.

[13] G. L. Gaines, *Insoluble Monolayers at Liquid-Gas Interfaces*, Science, New York (1965).

[14] Y. Sato, C. X. Wu, Y. Majima, and M. Iwamoto, *Jpn. J. Appl. Phys.*, **37** (1998) 5655; *J. Colloid Interface Sci.*, **218** (1999) 118.

[15] C. Bucci and R. Fieschi, *Phys. Rev.*, **148** (1966) 816.

[16] J. G. Parsons, *Phys. Rev. Lett.*, **41** (1978) 877.

[17] Y. Majima, Y. Sato, and M. Iwamoto, *Jpn. J. Appl. Phys.*, **36** (1997) 5237.

[18] T. Hino, *J. Appl. Phys.*, **48** (1976) 816; T. Hino, *Trans. IEEE*, **EI-21** (1986) 1007; T. Hino, *J. Appl. Phys.*, **46** (1975) 1956.

[19] R. Chen and Y. Kirsh, *Analysis to Thermally Stimulated Process*, Pergamon, Oxford (1981).

[20] K. S. Lee and M. Iwamoto, *Mol. Cryst. Liq. Cryst.* (2000) in press.

CHAPTER 6

CHIRAL PHASE SEPARATION

In nature, the combination of two molecular enantiomers of left- and right-handed chirality to form an racemic dimer is so abundant. Thus, means of separation of the dimers into distinct phases of optical activity, the chiral phase separation (CPS), are of critical importance in physics, chemistry, biology, and geology, [1] and have long been known in experiment, going back to Pasteur in 1848 in isolation of enantiomers of sodium tartrate by crystallization [2] and to Jungfleish in 1882 in the localization of crystallization of individual enantiomers on suitably disposed seeds. [3] However, the major challenge in the study is in the theoretical side to understand the mechanism of the chiral symmetry breaking (CSB) in CPS. The study of CPS in three-dimensional (3D) state is especially difficult because both left- and right-handed chiral molecules have been classically regarded as exact mirror images and, therefore, are energetically equivalent in bulk due to the molecular rotation. It is natural to think that this difficulty may be reduced in a monolayer of racemic amphiphiles at air/water interface because the molecular rotation of the amphiphiles is limited due to the hydrophobic interaction between the amphiphiles and the water. In fact, a remarkable progress of the study in a 2D system was performed by Lundquist in 1960s. [4, 5] Recognizing the importance of steric factors in the living system, such as in the architecture of the lipoprotein membranes of cells, Lundquist found that at a temperature below 20°C the racemic monolayer of the methyl ester of 2-methylhexacosanoic acid shows a pressure plateau in each pressure-area isotherm. This was interpreted as a 2D phase transition region, from the expanded phase to the condensed 2D crystalline phase. At that time, however, there were no methods available for the direct crystallographic study for the crystalline phase. Until 1993, a significant advance was obtained by Eckhardt et al. [6] By transferring such a monolayer of each of surface pressure from water surface onto mica and using atomic-force microscopy, they detected that the phase transition at the region of small molecular area is a CPS transition. More specifically, a transition from rectangular lattices of molecular parking structure (racemic phase) into a chiral phase in which the enantiomers form two antipodal oblique lattices distinguished by some stripes (see results of Eckhardt [6]).

In physics, the CSB and stripe formation have been extensively studied in hexatic LC films, [7] monolayers, [8, 9] smectic films, [9] and thin ferroelectric LC. [10, 11] These papers show that the phase separation of enantiomers is not the only mechanism leading to CSB and stripe formation, e. g. breaking up-down symmetry also leads to the same qualitative behavior. [10, 11] In theoretical approach presented by Selinger et al., [9] the CSB and formation of stripe and other patterns are described with the

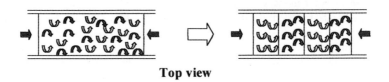

Figure 6.1 Compression-induced chiral phase separation.

help of the Landau free energy expansion in terms of the chiral order parameter. Hence, it serves to describe the temperature-effect in CSB. So far, however, there is no approach to the investigation connecting the CPS and 2D phase transition in racemic monolayers by compression. (Figure 6.1 shows a typical CPS transition of monolayer from mixed chiral enantiomers to an one-dimensional periodical chiral pattern by compression.) The theory of CSB or CPS associated with stripe-pattern formation induced by surface-pressure, [4, 5, 6] therefore, becomes an urgent task. In this chapter, using a continuum theory, the surface-pressure-induced CPS in the racemic monolayers is discussed.

6.1 Elastic Energy and Bragg-Williams Mixing Energy

The approach in this chapter follows the work by Ou-Yang and Liu, dealing with tilted chiral lipid membranes [12, 13] by viewing a monolayer as a film of cholesteric liquid crystal (ChLC). [14] However, to describe the racemic property, the ChLC is considered as a mixture of the left- and right-handed chiral enantiomers with same molecular number. Instead of using a linear chiral curvature elastic energy term as done by Ou-Yang *et al.*, the complete 3D-director elastic free energy density of a ChLC [15]

$$g_{LC} = \frac{1}{2}[k_{11}(\nabla \cdot \mathbf{d})^2 + k_{22}(\mathbf{d} \cdot \nabla \times \mathbf{d} - \frac{k_2}{k_{22}})^2 + k_{33}(\mathbf{d} \times \nabla \times \mathbf{d})^2] \quad (6.1)$$

is introduced to describe the free energy of the molecular orientation in monolayers. Here, $\mathbf{d}(= \sin\theta\cos\phi, \sin\theta\sin\phi, \cos\theta)$ is the 3D-director field and k_{11}, k_{22}, and k_{33} are the splay, twist, and bend elastic constants, respectively. However, the chiral modulus k_2 now is naturally defined as

$$k_2 = k_{20}(2\chi(\mathbf{r}) - 1), \quad \chi = N_L(\mathbf{r})/[N_L(\mathbf{r}) + N_R(\mathbf{r})], \quad (6.2)$$

where $k_{20} > 0$ is the chiral elastic constant for pure left-handed chiral phase, $N_L(\mathbf{r})$ and $N_R(\mathbf{r})$ are the local densities of the left- and right-handed chiral enantiomers, respectively, and $\chi(\mathbf{r})$ is obviously the chiral order parameter that characterizes the CPS: (i) left-handed chiral phase for $1 \geq \chi > 1/2$; (ii) racemic phase for $\chi = 1/2$; and (iii) right-handed chiral one for $1/2 > \chi \geq 0$. With the improvement in description, the CSB can now be dealt dependent not only on temperature T and the chirality (k_{20}) but also on the surface-pressure Π and molecular area A_0. Because in the discussed phase transition region the monolayers have been known as LC or crystalline, [4, 5, 6] the incompressibility in volume makes the angle θ, the angle between the director $\mathbf{d}$ and the normal of the monolayers, uniform as (see Fig. 6.2)

$$\cos\theta = V_0/lA_0, \quad (6.3)$$

where l is the average molecular length and V_0 is the average molecular volume of the amphiphiles. Both are constant. In other words, the chiral order parameter χ, the orientation pattern $\phi(\mathbf{r})$, and the molecular area A_0 (which is relating to surface pressure Π) have together been incorporated into the elastic free energy density g_{LC}. Another important step in the present theory is that the additional free energy of the mixing effect by Bragg-Williams approach in the study of binary alloy is also considered [16, 17]

$$\begin{aligned} F_{MIX} &= \frac{1}{A_0}\int[-\frac{z}{2}w_{RR} - \frac{z}{2}(w_{LL} - w_{RR})\chi + zw\chi(1-\chi) \\ &+ k_BT\{\chi\ln\chi + (1-\chi)\ln(1-\chi)\}]dA, \quad (6.4) \end{aligned}$$

where dA/A_0 gives the local density of molecules, k_B is the Boltzmann constant, $w = (1/2)(w_{LL} + w_{RR}) - w_{LR}$ with $-w_{\alpha\beta}$ (α, β = L, R) denoting the nearest-neighbor interaction energy between α-handed enantiomer and β-handed one, and z is the number of nearest-neighbor of one molecule. Obviously, the last term in the integrand associated with k_BT is the entropy-increase on mixing, and the other terms relating to $w_{\alpha\beta}$ is the enthalpy of mixing calculated by the renowned Bragg-Williams approximation. For the experiment discussed by Eckhardt *et al.*, [6] z has been found to be equal to 4 for 2D rectangular and oblique lattices and $w_{LL} = w_{RR}$. Hence, by neglecting the first constant term Eq. (6.4) can be reduced to

$$F_{MIX} = \frac{k_B T}{A_0} \int [\chi \ln \chi + (1 - \chi) \ln(1 - \chi) + \alpha_0 \chi(1 - \chi)] dA, \qquad (6.5)$$

where $\alpha_0 = 4w/k_B T$. As definition for w, the positive α_0 favors CPS and it consists of short range repulsive interaction, which would be increasing with compression. For example, as w approaches ∞, F_{MIX} takes its minimum at $\chi = 0, 1$, the complete CPS state. On the other hand, the entropy favors racemic phase, e. g. , let $T \to \infty$, then $\chi = 1/2$ minimizing F_{MIX}. Before getting into actual calculation, therefore, it can be seen that the CPS transition being just the result of competition between the entropy (temperature T) and enthalpy (compression, or molecular area A_0). To understand it quantitatively, the curvature elastic energy must be combined with the Bragg-Williams mixing energy as the free energy for the binary ChLCs film as

$$F = l \cos \theta \int g_{LC} dA + F_{MIX} + \lambda \int (\chi - \frac{1}{2}) dA, \qquad (6.6)$$

where λ is the Lagrange-multiplier associated with the condition of $\int (\chi - 1/2) dA = 0$, a condition based on the fact that the initial phase before compression is a racemic phase, i. e. $\int N_L dA = \int N_R dA$. The equilibrium equations for both $\phi(\mathbf{r})$ and $\chi(\mathbf{r})$ are obtained by minimizing the free energy. These equations and their solutions show that the CSB or CPS is associated with the complex localization of individual enantionmers in the monolayers, the CPS pattern formation dependent on T and A_0. It is shown that below a certain critical temperature compression can induce phase transition from racemic phase to CPS phase with a stripe-pattern formation. This gives a clear and direct explanation of CPS in the monolayers of racemic amphiphiles observed by Lundquist [4, 5] and Eckhardt *et al.* [6]

6.2 Chiral Phase Separation

6.2.1 One-dimensional Periodical Solution

In the actual calculation, a Cartesian coordinate xyz is taken as illustrated in Fig. 6.2, where the plane of $z = 0$ is the water/air interface. For convenience, the one-constant approximation as $k_{11} = k_{22} = k_{33} = k$ is used. The curvature energy density Eq. (6.1) in terms of ϕ is

$$g_{LC} = \frac{1}{2}[k \sin^2 \theta(\phi_x^2 + \phi_y^2) - k_2 \sin 2\theta(\cos \phi \phi_x + \sin \phi \phi_y) + k_2^2/k], \qquad (6.7)$$

where $\phi_x = \partial \phi / \partial x$ and $\phi_y = \partial \phi / \partial y$. Since the free energy F is a function of $\phi(x, y)$ and $\chi(x, y)$, two Euler-Lagrange (EL) equations can be derived from the variation of

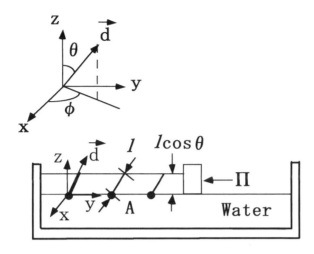

Figure 6.2 Geometry of the monolayer of racemic amphiphiles at the air-water interface. The thickness of the monolayer is $l \cos \theta$, where l is the average length of the molecules along their long axis and θ is the average value of the angle between the molecular long axis and the normal of the air-water interface. The total bulk volume of the monolayer is $V = lA \cos \theta = NlA_0 \cos \theta$, where N is the molecular number, A the total area, and A_0 the molecular area. Therefore, one has relation of $\cos \theta = V_0/lA_0$ where $V_0 = V/N$.

F with respect to ϕ and χ (see Appendix). The first EL equation is mainly related to ϕ as

$$\Delta \phi = \phi_{xx} + \phi_{yy} = \frac{1}{k} \cot \theta [k_{2,x} \cos \phi + k_{2,y} \sin \phi], \qquad (6.8)$$

obtained by $\delta F/\delta \phi = 0$, where Δ is a 2D-Laplace operator, $k_{2,x} = \partial k_2/\partial x$, and $k_{2,y} = \partial k_2/\partial y$. Note that in pure racemic phase ($k_2 = 0$) and pure chiral phases ($k_2 - k_{20}$ or $k_2 = -k_{20}$), Eq. (6.8) is identical to the 2D-pattern equation in usual LC, $\Delta \phi = 0$. [15]

The second EL equation derived from $\delta F/\delta \chi = 0$ is given by

$$\lambda + \frac{k_B T}{A_0} [\ln \frac{\chi}{1 - \chi} + \alpha_0 (1 - 2\chi)] + k_{20} l \cos \theta [\frac{2k_2}{k} - \sin 2\theta (\phi_x \cos \phi + \phi_y \sin \phi)] = 0. \quad (6.9)$$

Interestingly, at high temperature limit, i.e., as $T \to \infty$, the equation predicts an obvious solution of $\chi = 1/2$, which is nothing but a pure racemic phase, i.e., an isotropic phase. This is a reasonable result in LC physics, [15] and consistent with the observation by Lundquist that the CPS transition for a racemic mixture must occur below some critical temperature. [4, 5]

In principle, Eqs. (6.8) and (6.9) associated with $\int(\chi - 1/2)dA = 0$ give complete conditions for determining λ, ϕ, and χ at any T and A_0. However, it is quite a task to find general solutions for Eqs. (6.8) and (6.9) analytically owing to their high-order nonlinearity. Fortunately, the focus of the discussion is on the phase transition region, the proximity of $\chi = 1/2$. Therefore, using $\ln[\chi/(1 - \chi)] \simeq 4(\chi - 1/2)$, Eq. (6.9) becomes

$$k_2^* = \eta k_{20}(2\chi - 1) = \frac{k}{2}[\sin 2\theta(\phi_x \cos \phi + \phi_y \sin \phi) - \frac{\lambda}{lk_{20} \cos \theta}], \qquad (6.10)$$

where $\eta = 1 + k(2 - \alpha_0)k_BT/2V_0k_{20}^2$. Taking its partial derivatives with respect to x and y, respectively, gives

$$\chi_x = \frac{k \sin 2\theta}{4\eta k_{20}}[\phi_x \cos \phi + \phi_y \sin \phi]_x, \quad \chi_y = \frac{k \sin 2\theta}{4\eta k_{20}}[\phi_x \cos \phi + \phi_y \sin \phi]_y, \qquad (6.11)$$

where $\chi_x = \partial \chi/\partial x$, $[...]_x = \partial[...]/\partial x$, and so on. Substituting Eq. (6.11) into Eq. (6.8) results in the following equation for ϕ:

$$\Delta\phi = \eta^{-1}\cos^2\theta[\psi_x \cos \phi + \psi_y \sin \phi], \qquad (6.12)$$

where $\psi = \phi_x \cos \phi + \phi_y \sin \phi$, $\psi_x = \partial \psi/\partial x$, and $\psi_y = \partial \psi/\partial y$. It can be seen that for strong chirality, i.e., $k/k_{20} = p_0/2\pi \to 0$ where p_0 is the pitch of the equilibrium 3D-helical structure of the corresponding ChLC, [16, 17] $\eta \to 1$. In the following discussion, it is assumed that $\eta > 0$ being always held to emphasize the effect of strong chirality.

In general, Eqs. (6.10) and (6.12) associated with Eq. (6.3) predict that the compression of monolayers can induce the CPS and the corresponding formation of complex orientation pattern with either 1D or 2D structures. Some implicit and analytic 2D solutions of Eq. (6.12) may be possible to find. However, for simplicity and for comparison with the experiment (Fig. 4 in Ref. [6]), only the 1D case is considered. Letting $\phi_y = 0$ and $\psi_y = 0$, Eq. (6.12) can be transformed to an ordinary differential equation as

$$\frac{d^2\phi}{dx^2} = \cos^2\theta^*[\cos \phi \frac{d}{dx} \cos \phi \frac{d}{dx}\phi], \qquad (6.13)$$

where $\cos^2\theta^* = \eta^{-1}\cos^2\theta$. The solutions of Eq. (6.13) represent the possible 1D patterns which are results by minimizing free energy F given in Eq. (6.6). First, there is the solution of $\phi = \phi_0$ where ϕ_0 is a constant. This solution corresponds to the racemic phase, i.e., the stripe of infinite period with $F = (N_L + N_R)(\ln \frac{1}{2} + \frac{1}{4}\alpha_0)k_BT$, regarded as the initial phase. Second, the implicit and general solutions which show striped pattern parallel to y-axis with finite period, i. e. the CPS phase, (see Appendix) can be derived as:

$$x - x_0 = (\frac{t}{4})E(\phi(x) - \frac{\pi}{2}, \cos \theta^*)/E(\frac{\pi}{2}, \cos \theta^*), \qquad (6.14)$$

where x_0 and t are two integral constants, the latter is also the period of the stripe; and

$$E(\phi, \cos\theta^*) = \int_0^\phi \sqrt{1 - \cos^2\theta^* \sin^2\phi'}d\phi' \tag{6.15}$$

is a second ellipse integral. With Eq. (6.14) and $\int(\chi - 1/2)dA = 0$, it is obtained from Eq. (6.10) $\lambda = 0$ and the effective chiral order parameter of the CPS pattern, or localization of enantionmers, corresponds to the stripe

$$ch(x) \equiv \chi(x) - \frac{1}{2} = \frac{p_0 \sin 2\theta \cos\phi(x) E(\pi/2, \cos\theta^*)}{2\pi\eta t\sqrt{1 - \cos^2\theta^* \cos^2\phi(x)}}, \tag{6.16}$$

as shown schematically in Fig. 6.3. Chiral separation becomes stronger as θ decreases,

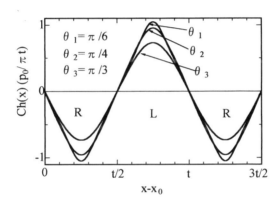

Figure 6.3 The θ dependence of the effective chiral order parameter $ch(x) = \chi - 1/2$ which characterizes the chirality (positive for left-handed chiral phase and negative for right-handed one) and the strength of the chiral phase separation. The unit of $\chi - 1/2$ used in the calculation if p_0/t where p_0 is the pitch of the equilibrium helical structure for the pure chiral phase in bulk and t is the period of the stripe. The calculation is based on Eq. (6.16) with the strong chirality case of $\eta \to 1$.

indicating that the separation is just a result of monolayer compression. Substituting the stripe solution Eq. (6.14) into Eq. (6.1), it can be shown the elastic energy [the first integral in Eq. (6.6)] in an area of $x \times y = t \times L$ to be (see Appendix)

$$F_C = 8kLl\cos\theta\sin^2\theta[(\eta^{-1} - 1)K^*E^* + (2 - \eta^{-1})E^{*2}]/t, \tag{6.17}$$

and dividing by the molecular number of the area, $N = l \cos \theta t L / V_0$ (see Fig. 6.1), yields the elastic energy per molecule

$$f_C = F_C/N = 8kV_0 \sin^2 \theta [(\eta^{-1} - 1)K^* E^* + (2 - \eta^{-1})E^{*2}]/t^2. \qquad (6.18)$$

Here, $E^* = E(\pi/2, \cos \theta^*)$ is the complete second ellipse integral and K^* is the complete first ellipse integral as

$$K^* = \int_0^{\pi/2} d\phi / \sqrt{1 - \cos^2 \theta^* \sin^2 \phi}. \qquad (6.19)$$

In the proximity of racemic to CPS transition the integrand of Eq. (6.5) can be expanded in power of $(\chi - 1/2)$ to yield

$$F_{MIX} = \frac{k_B T}{A_0} \int [\ln \frac{1}{2} + \frac{1}{4}\alpha_0 + (2 - \alpha_0)(\chi - \frac{1}{2})^2 + \frac{4}{3}(\chi - \frac{1}{2})^4] dA. \qquad (6.20)$$

Substituting Eq. (6.14) and Eq. (6.16) into Eq. (6.20) and dividing by the molecular number N, with a lengthy calculation (see Appendix), the Bragg-Williams mixing energy per molecule is obtained

$$
\begin{aligned}
f_{MIX} = {} & k_B T[\ln \frac{1}{2} + \frac{1}{4}\alpha_0 + (2 - \alpha_0)\eta^{-1}(E^* K^* - E^{*2})(\frac{p_0 \sin \theta}{\pi t})^2 \\
& + \frac{4}{3}\eta^{-2}(\sin^{-2} \theta^* E^{*4} - 2K^* E^{*3} + E^{*4})(\frac{p_0 \sin \theta}{\pi t})^4].
\end{aligned}
\qquad (6.21)
$$

Hence, the free energy per molecule can be expressed as a quadratic function of $q^2 = (p_0 \sin \theta / \pi t)^2$ as

$$f = f_C + f_{MIX} = a_0 - a_2 q^2 + a_4 q^4, \qquad (6.22)$$

with $a_0 = (1/4)k_B T(\alpha_0 - 4 \ln 2)$, $a_2 = (\alpha_0 - 2)(E^* K^* - E^{*2})\eta^{-1}k_B T - 8kV_0(\pi/p_0)^2[(2 - \eta^{-1})E^{*2} + (\eta^{-1} - 1)K^* E^*]$, and $a_4 = (4/3)(E^{*4}/\sin^2 \theta^* - 2K^* E^{*3} + E^{*4})\eta^{-2}k_B T$. It is clear that racemic to CPS transition occurs under the necessary condition $a_2 > 0$ (see Fig. 4). This requires that at least $\alpha_0 > 2$. From the definition of $\alpha_0 = 4w/k_B T$ the critical temperature is obtained

$$T_c = 2w/k_B = [(w_{LL} + w_{RR}) - 2w_{LR}]/k_B, \qquad (6.23)$$

below which the racemic to CPS phase transition can happen. This is what was observed by Lundquist. [4, 5] The critical period of the CPS structure (t_c, at which the free energy of the CPS phase is equal to that of racemic one, i.e., $t \to \infty$) can be obtained as $t_c = (p_0 \sin \theta / \pi)\sqrt{a_4/a_2}$. Performing minimization of f with t gives the equilibrium period of $t_m = (1/2)t_c$. Obviously, existence of definite t_m needs again $a_2 > 0$, at the case of strong chirality (i.e., $\eta \to 1$ and $\theta^* \to \theta$, thence, E^* and $K^* \to E(\cos \theta)$ and $K(\cos \theta)$, respectively), $[K(\cos \theta)/E(\cos \theta) - 1] > 8kV_0(\pi/p_0)^2[4w -$

$2k_BT]$ is necessary. From the property of both the complete ellipse integrals of K and E, the inequality reveals that θ must be less than a critical angle of θ_c given by

$$K(\cos\theta_c)/E(\cos\theta_c) = 1 + 8kV_0(\frac{\pi}{p_0})^2/(4w - 2k_BT). \tag{6.24}$$

Accordingly, from Eq. (6.3), CPS occurs at the following condition

$$A_0 < A_c = V_0/l\cos\theta_c. \tag{6.25}$$

It is then found that compression plays a critical important role for CPS with stripe-pattern formation.

6.2.2 One-dimensional Chiral Phase Separation

In a quantitative sense the present theory can be understood by comparing with the experiment and calculation by Eckhardt *et al.* [6] It is found that the energies of the oblique (CPS phase for $\chi = 0, 1$) and rectangular (racemic phase with $\chi = 1/2$) differ by ~ 7.7 kJ Mole^{-1} with the former being more stable. From the Bragg-Williams formula Eq. (6.5), the difference of free energies per molecule between the racemic and complete CPS phases $f(\chi = 1/2) - f(\chi = 0, 1) = k_BT[\ln(1/2) + \alpha_0/4] = 7.7 \times 10^3 \times 10^7$ erg$/6 \times 10^{23} = 1.3 \times 10^{-13}$ erg can be obtained. Taking $k_BT = 4 \times 10^{-14}$ erg at $T = 20°C$ yields $\alpha_0 = 14.2$, which satisfies $\alpha_0 > 2$, the request of the critical temperature condition Eq. (6.23). From the value of α_0 we can estimate the difference of the nearest-neighbor interaction energies $w = (1/2)(w_{LL} + w_{RR}) - w_{LR} = (1/4)\alpha_0 k_BT = 1.42 \times 10^{-13}$erg which is comparable very well with the membrane (bilayer of amphiphiles) curvature modulus $2lk \sim 10^{-13}$ erg obtained by Helfrich by viewing the membranes as LC films. [18] It is also important to note that the present assumption of strong chirality is invalid for usual thermotropic ChLC with pitch of order of 1 μm [15] because inserting the order of pitch, the above estimate of $\alpha_0 = 14.2$, and the typical values $k = 10^{-6}$ erg/cm, $V_0 = 10^{-21}$ cm^3, $T = 300$ K, into the formula of η [defined after Eq. (6.10)] gives the parameter of $\eta - 1$ to be of order of 10^3. This is in strong contradiction with the assumption $\eta - 1 << 1$ (or, at least, $\eta < 1$) used above. In other words, the present theory is valid only if the chirality is two orders of magnitude larger compared with usual thermotropic ChLC, i. e. with pitch of order of 0.01 μm. This is quite possible in the case of chiral amphiphiles observed by Eckhardt and co-workers. [6] Ou-Yang *et al.* showed that the pitch in the bilayer of tilted chiral amphiphiles viewed as ChLC can be determined by the formula $p_0 = p_{ch} = 2\pi\rho_0\sin^2\theta/(1 + \sin^2\theta)$. [12, 13] Here, ρ_0 is the radius of the helical tubes formed by the tilted chiral bilayer [19] and θ is the tilt angle of the hydrocarbon chains of the amphiphiles from the normal of the surface of the bilayer. The order of diameter of the tubes is 0.5 μm [19] and tubes of diameter of 0.05 μm are also found. [20] Therefore, if $\sin\theta$ is about 0.1 then p_0 will be approximately 0.01 μm, as the chiral amphiphiles is confirmed by the tube formation. [19] This shows the validity of the model for the observation reported by Eckhardt [6] because the pitch of the chiral amphiphiles in the monolayer should be same as in their bilayer. These confirm the reasonability of present theory: The CPS of monolayer of amphiphiles and its associated pattern formation can be discussed well as a film of Bragg-Williams binary cholesterics of the left- and right-handed chiral enantiomers.

It is important to point out that the irrelevance between dipole moment and chirality excludes the possible influence of electrostatic interaction on the CPS phase transition. It is reasonable to predict [21, 22] that the additional surface pressure due to the CPS phase transition can be calculated by the formula $\delta\Pi = -\partial f/\partial A_0$, which appears a function form of $-b_3 A_0^{-3} + b_5 A_0^{-5} + ...$ with $b_3, b_5 > 0$ and adds into the typical pressure-area isotherms of 2D van der Walls equation of state. [23] This is the mechanism of the pressure plateau in the area-pressure isotherms associated with the CPS transition. [4, 5, 6]

Finally, the connection between the present theory and those given by Selinger *et al.* [9] They expressed the elastic energy as a function of gradients of 2D-director $\mathbf{c} = (\cos\phi, \sin\phi, 0)$, which is parallel to the project direction of molecular director $\mathbf{d}$ in xy-plane [see Eq. (1) in Ref. [9]]. With a lengthy calculation, it was shown that the elastic energy given in Eq. (6.1) can be also expressed with the function of the gradients of $\mathbf{c}$ as

$$
\begin{aligned}
g_{LC} = \;& \frac{1}{2}k_{11}\sin^2\theta(\nabla'\cdot\mathbf{c})^2 \\
+ \;& [\frac{1}{2}k_{22}\sin^2\theta\cos^2\theta + \frac{1}{2}k_{33}(\sin^4\theta - 4\sin^2\theta)](\mathbf{z}\cdot\nabla'\times\mathbf{c})^2 \\
+ \;& 2k_{33}\sin^2\theta(\nabla'\times\mathbf{c})^2 - k_2\sin\theta\cos\theta\mathbf{z}\cdot\nabla'\times\mathbf{c}) \\
- \;& 2k_2\sin^2\theta(\mathbf{c}\cdot\nabla'\times\mathbf{c}) \\
+ \;& 2(k_{22} - k_{33})\sin^3\theta\cos\theta(\mathbf{c}\cdot\nabla'\times\mathbf{c})(\mathbf{z}\cdot\nabla'\times\mathbf{c}) \\
+ \;& 2(k_{22} - k_{33})\sin^4\theta(\mathbf{c}\nabla'\times\mathbf{c})^2 + \frac{1}{2}k_2^2/k_{22}\;,
\end{aligned}
\tag{6.26}
$$

where $\nabla' = (\partial_x, \partial_y, 0)$ and $\mathbf{z} = (0, 0, 1)$ being the normal vector of the monolayer. Obviously, this formula includes all possible combinations of linear and quadratic gradients of $\mathbf{c}$ is more complete than those used by Selinger and Hinshaw. [9, 10, 11] However, it is same that the terms of the gradients of the chiral order parameters are taken into account in the theoretical treatment as shown in Eq. (6.11) and those by Selinger and Hinshaw.

6.3 Discrete One-dimensional CPS Solution

6.3.1 One-dimensional General Solution

Now a special case will be discussed: the discrete one-dimensional chiral phase separation. [24] Actually, an 1D general solution to Eqs. (6.8) and (6.9) can be found and written, without any approximations, in a form of

$$\cos\phi = \frac{\sqrt{k_B T k}}{k_{20}\cos\theta\sqrt{8V_0}}\frac{G'(\chi)}{\sqrt{G(\chi)}}, \tag{6.27}$$

$$x - x_0 = \frac{k\tan\theta}{8k_{20}}\int^\chi \frac{[G'(\chi')]^2 - 2G(\chi')G''(\chi')}{[G(\chi')]^{3/2}\{G(\chi') - \frac{k_B T k}{8V_0 k_{20}^2\cos^2\theta}[G'(\chi')]^2\}^{1/2}}d\chi', \tag{6.28}$$

where $G(\chi) = C + \chi\ln\chi + (1-\chi)\ln(1-\chi) + \alpha_0\chi(1-\chi)$, with C an integral constant, and $G'(\chi)$ and $G''(\chi)$ are the first and second derivatives, respectively. Given this, it is immediately found that Eqs. (6.15) and (6.16) are nothing but the solutions under a second-order approximation $G(\chi) \approx C - \ln 2 + \alpha_0/4 + (2 - \alpha_0)(\chi - 1/2)^2$. Noting that this approximation stands only when $|\chi - 1/2|$ is small, it is readily understood that the solution Eq. (6.16) is, at most, a "weak" CPS.

6.3.2 One-dimensional Discrete Pattern

As shown in Fig. 6.4, as $\alpha_0 > 2$, the mixing energy has two minimums at $\chi = \chi_1$ or χ_2. Noting that the energy difference between the racemic state ($\chi \equiv 1/2$) and the CPS state $\chi = \chi_1$ or χ_2 increases linearly with respect to α_0, it is reasonable to expect that as $\alpha_0 \gg 2$, if all domains of the monolayer have chiral order parameters $\chi = \chi_1$ or χ_2, the mixing energy would be greatly decreased. The fact that the mixing energy Eq. (6.4) does not include a concentration square gradient (CSG) term, or $\nabla\chi$-related terms, does not restrain the sudden variation of the chirality χ with respect to position. This allows the sudden change of the chirality χ from one domain to its neighbors. With the above considerations, it can be reasonably conjectured that it is energetically advantageous that, in case of $\alpha_0 \gg 2$, the monolayer would rather develop an optimal pattern consisting of several domains, each of which has homogeneous chiral order parameter of either χ_1 or χ_2. It is easy to find that such a pattern corresponds to an extreme case of Eqs. (6.27) and (6.28), with the integral constant $C = -(\chi_1\ln\chi_1 + \chi_2\ln\chi_2 + \alpha_0\chi_1\chi_2)$. Put the other way around, a discrete one-dimensional solution is allowed. As C equals to this special value, Eqs. (6.27) and (6.28) are simplified into a series of stripes $\cdots L_{-2}, L_{-1}, L_0, L_1, L_2 \cdots$ schematically shown in Fig. 6.5 with

$$\phi \equiv \pi/2, \qquad \chi = \begin{cases} \chi_1, & x \in \cdots L_{-1}, L_1 \cdots, \\ \chi_2, & x \in \cdots L_{-2}, L_0, L_2 \cdots. \end{cases} \tag{6.29}$$

This solution predicts that within one single stripe the azimuthal angle ϕ is constant. This is distinctively different from the prediction by Ginzburg-Landau-type theory, [9] in which the azimuth ϕ linearly varies with respect to the 2D coordinate. This offers a possible explanation of the homogeneous bond orientation within each stripe observed by Eckhardt *et al.* The fast variation of the chirality χ at the borders of adjacent domains implied by Eq. (6.29) is also supported by the image reported by Eckhardt

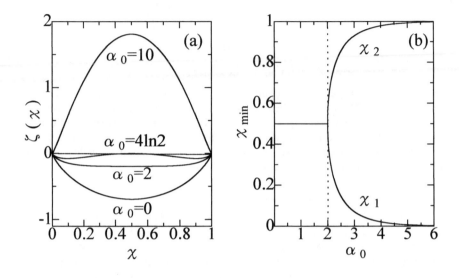

Figure 6.4 (a) Behavior of $\zeta(\chi) = \chi \ln \chi + (1 - \chi) \ln(1 - \chi) + \alpha_0 \chi(1 - \chi)$ with several values of chirality discrimination coefficient α_0. (b) χ_{min}, the values of χ which minimize the function $\zeta(\chi)$, as a function of α_0. As $\alpha_0 = 2$, there is a ramification of χ_{min}; as α_0 further increases, the two values of χ_{min}, denoted as χ_1 and χ_2, quickly approach 0 and 1 respectively.

et al., in which the sharp mutation of the bond orientation is evident, with only one or two lines of intermediate molecules. On the other hand, since within one single domain both ϕ and χ are constant, the free energy Eq. (6.6) does not define the widths of the stripes. The Lagrange condition $\int(\chi - 1/2)dx = 0$ loosely requires that the total widths of L and R bands be identical, i.e., $\sum\limits_{i=odd} L_i = \sum\limits_{i=even} L_i$, whereas the widths of single stripes still remain quite free. This uncertainty gives an elementary explanation why the widths of stripes are so different. [6] It should be attributed to the process of CPS which is quite uncertain. This point will be further discussed in the following sections.

6.3.3 Line Tension

Now attention will be focused on the border of two neighboring stripes with chiralities χ_1 and χ_2, respectively. In Ginzburg-Landau-type theory, the CSG term will lead to an effective line tension at the edge. In the present theory, the fast variation of chirality with position implies that the border is similar to the disclination in liquid crystals, which leads to additional energy corresponding to the line tension. As shown in Fig. 6.5, the interaction between the two sides of a border is different

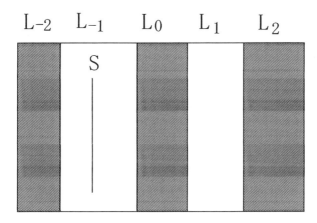

Figure 6.5 Schematic of striped pattern of strong CPS in two dimensions.

from that between two sides of a line in a homogeneous domain (such as line S schematically shown in stripe L_{-1}) due to CD effect. Considering only the nearest-neighbor interaction, the difference between these two kinds of interactions in unit length, i.e., the line tension at the edge, $\lambda_l = 2\rho(\chi_2 - \chi_1)^2 w$, where ρ is the molecular line density along the edge is simply estimated. As CPS occurs, $\chi_1 \neq 1/2$, it always stands that $\lambda_l > 0$. As a remark, this line energy is a part of the mixing energy, whereas it is not included in the continuum approach by Bragg and Williams (similar to the line tension of a disclination in liquid crystals which is not included in the continuum elastic energy), since the local chiral order parameter χ is used to describe the monolayer, a condition has been implicitly assumed that the monolayer can be viewed as a microscopically homogeneous system. Consequently, in the case that the chirality varies quickly with position, this assumption will be invalid, and a line tension is directly yielded from the intermolecular CD effect. Fortunately, even with consideration of the line tension, Eq. (6.29) is still a stationary solution to the model.

6.3.4 Pattern Formation

The presence of the line tension indicates that the longer is the total length of borders, the higher is the whole energy. This leads to a natural conclusion that the minimal-energy solution is a pattern consisting of two large domains with one border between them. All molecules separate into the two domains with a left-handed $\chi = \chi_1$ and a right-handed $\chi = \chi_2$ (this is a special case of Eq. (6.29)). However, this rarely happens. In fact, the multi-stripe patterns are metastable, namely, if two stripes with identical chirality, such as L_0 and L_2 in Fig. 6.5, converge with each other, the total energy will be lowered. The metastablity of this pattern may be closely related to the 2D nature of the Langmuir monolayer. Saying it more clearly, as CPS happens in a racemic monolayer, the two kinds of enantiomers have to, at first, collect at small domains. Consequently, it is possible for the monolayer to reach a pattern consisting of a series of stripes. Once a multi-stripe pattern happens, it would be quite difficult to evolve into the two-domain pattern through a 2D permeation. In other words, although the two-domain pattern is energetically advantageous, for a large-area monolayer it is quite difficult to reach. Now it can be recognized that the multi-stripe pattern may be very stable, and what kind of pattern happens should be determined by the process of CPS.

Consider the regime of $\alpha_0 \sim 2$, which corresponds to the beginning of CPS. A perturbation analysis indicates that, for striped pattern $\phi = \phi_0 \sin qx$, $\chi = 1/2 + \chi_0 \cos qx$, the CSB begins when $\alpha_0 = 2 - \Delta\alpha_0$, with $\Delta\alpha_0 = k_{20}^2 V_0 \cos^2 \theta / 2k_B Tk$. Whereas for the pattern $\phi = \pi/2 + \phi_0 \cos qx$, $\chi = 1/2 + \chi_0 \cos qx$, which corresponds to Eqs. (4) and (5), the threshold is $\alpha_0 = 2$. This supports a proposition that the difference between these two thresholds is not important. As a simple evaluation, by using the typical data of liquid crystals $\Delta\alpha_0 \sim 10^{-6}$ is obtained. It is completely negligible in comparison with the variation range of α_0 during compression, which is on a scale of 10 in Eckhardt's experiment, [6] as pointed out in section 6.2. So attention may not be paid to the details of this regime, but focus on the behavior of strong CPS region with large α_0. Noting that the CD coefficient α_0 must vary with the intermolecular distance, and in the case of CPS it should increase as molecular area decreases, the picture of compression-induced CPS is qualitatively depicted as follows. As average molecular area A_0 is large enough, α_0 is small and the monolayer is racemic. Further compression increases α_0 to the threshold of (about) 2. Then, microscopically, two kinds of enantiomers begin to collect, respectively. Further compression successively enhances α_0 and at last leads to a strong CPS depicted by Eq. (6.29). In this process, shortwave patterns may happen at the beginning, since the two kinds of enantiomers have to collect, at first, at small range, respectively. However, shortwave modes are energetically disadvantageous because of the existence of the line energy mentioned earlier. Consequently the domains with identical chirality will converge with each other, if the distance among them is small enough. At last the monolayer reaches a pattern of parallel stripes with their widths larger than certain length scale (may be some 2D free running distance of molecules in Langmuir monolayer), and the whole system then becomes metastable. The AFM image discussed by Eckhardt *et al.* [6]

may be this metastable pattern. The above considerations also imply that the minimal width of stripes would be a length about several nanometers, with no relation to the 3D pitch of the chiral material. This is in good agreement with Eckhardt's experimental result, since the widths of the stripes are obviously several nanomaters.

6.4 Summary

The monolayer of racemic amphiphiles is studied as a film of cholesteric liquid crystal (ChLC) mixed by the left- and right-handed chiral enantiomers. The chiral phase separation (CPS) and molecular orientational pattern are analyzed by Bragg-Williams theory for binary mixtures and curvature-elastic model of ChLC. It is shown that below a certain critical temperature compression can induce transition from racemic phase to CPS phase with a stripe-pattern formation. This gives a clear explanation of CPS in a racemic monolayer recognized first by Lundquist [4] and observed recently by Eckhardt and co-workers. [6]

6.5 Appendix

6.5.1 Euler Differential Equation–Maxima and Minima of Definite Integrals

The necessary condition for the existence of either a maximum or a minimum of the definite integral [25]

$$I = \int_{x_1}^{x_2} F[y(x), y'(x), x] dx,$$ (6.30)

for fixed x_1 and x_2 is

$$
\begin{aligned}
\delta I &= \int_{x_1}^{x_2} \delta F dx \\
&\equiv \int_{x_1}^{x_2} [\frac{\partial F}{\partial y} \delta y + \frac{\partial F}{\partial y'} \delta y'] dx \\
&= \int_{x_1}^{x_2} [\frac{\partial F}{\partial y} \delta y dx + \frac{\partial F}{\partial y'} d\delta y] \\
&= \int_{x_1}^{x_2} [\frac{\partial F}{\partial y} \delta y - \frac{d}{dx} \left(\frac{\partial F}{\partial y'}\right) \delta y] dx + \frac{\partial F}{\partial y'} \delta y \mid_{x_1}^{x_2} \\
&= 0,
\end{aligned}
$$ (6.31)

for an arbitrary small variation δy. Hence every maximizing or minimizing function $y(x)$ must satisfy the differential equation

$$\frac{d}{dx} \left(\frac{\partial F}{\partial y'}\right) - \frac{\partial F}{\partial y} = 0$$ (6.32)

This is called *Euler's differential equation*. Here $y(x)$ must either assume given boundary values $y(x_1)$ and/or $y(x_2)$, or $y(x)$ must satisfy other conditions determining its boundary values.

Similarly, every set of n functions $y_1(x)$, $y_2(x)$, $\cdots$, $y_n(x)$ maximizing or minimizing the definite integral

$$I = \int_{x_1}^{x_2} F[y_1(x), y_2(x), \cdots, y_n(x); y'_1(x), y'_2(x), \cdots, y'_n(x); x] dx$$ (6.33)

must satisfy the set of n differential equations

$$\frac{d}{dx} \left(\frac{\partial F}{\partial y'_i}\right) - \frac{\partial F}{\partial y_i} = 0 \quad (i = 1, 2, ..., n)$$ (6.34)

together with suitably given boundary conditions. These are called Euler's differential equations. The Euler differential equations can also be extended to the case of multi-variables x_i $(i = 1, 2, ..., n)$

$$\sum_i \frac{d}{dx_i} \left(\frac{\partial F}{\partial y'_i}\right) - \frac{\partial F}{\partial y} = 0,$$ (6.35)

where y_i' refers to $\partial y/\partial x_i$. If F is a Lagrange function, then Eq. (6.32), Eqs. (6.34), and Eq. (6.35) are called *Euler-Lagrange differential equations*.

Using Euler-Lagrange differential equation Eq. (6.35), one immediately obtains

$$\frac{d}{dx}\left(\frac{\partial g_{LC}}{\partial \phi_x}\right) + \frac{d}{dy}\left(\frac{\partial g_{LC}}{\partial \phi_y}\right) - \frac{\partial g_{LC}}{\partial \phi} = 0 \tag{6.36}$$

from the condition that the free energy of the monolayer system Eq. (6.6) is minimum. With Eq. (6.7), the following can be obtained:

$$\frac{\partial g_{LC}}{\partial \phi_x} = k\sin^2\theta\phi_x - \frac{k_2}{2}\sin 2\theta\cos\phi$$

$$\frac{d}{dx}\left(\frac{\partial g_{LC}}{\partial \phi_x}\right) = k\sin^2\theta\phi_{xx} - \frac{k_{2,x}}{2}\sin 2\theta\cos\phi + \frac{k_2}{2}\sin 2\theta\sin\phi\phi_x \tag{6.37}$$

$$\frac{\partial g_{LC}}{\partial \phi_y} = k\sin^2\theta\phi_y - \frac{k_2}{2}\sin 2\theta\cos\phi$$

$$\frac{d}{dy}\left(\frac{\partial g_{LC}}{\partial \phi_y}\right) = k\sin^2\theta\phi_{yy} - \frac{k_{2,y}}{2}\sin 2\theta\cos\phi + \frac{k_2}{2}\sin 2\theta\sin\phi\phi_y \tag{6.38}$$

$$\frac{\partial g_{LC}}{\partial \phi} = \frac{k_2}{2}\sin 2\theta\sin\phi\phi_x - \frac{k_2}{2}\sin 2\theta\cos\phi\phi_y. \tag{6.39}$$

Substituting Eqs. (6.37), (6.38) and (6.39) into Eq. (6.36), one obtains Eq. (6.8).

6.5.2 Solution of Eq. (6.13)

Eq. (6.13) can be rewritten in another form

$$\frac{1}{2}\frac{d}{d\phi}\left(\frac{d\phi}{dx}\right)^2 = \frac{1}{2}\frac{d}{d\phi}\left[\cos^2\theta^*\cos^2\phi\left(\frac{d\phi}{dx}\right)\right]. \tag{6.40}$$

Suppose $x = x_0$ when $\phi = \pi/2$, the solution of the above equation becomes

$$x - x_0 = \left(\frac{t}{4}\right)E(\phi - \pi/2, \cos\theta^*)/E(\pi/2, \cos\theta^*), \tag{6.41}$$

where $E(\phi, \cos\theta^*)$ is defined by Eq. (6.15). This is nothing but Eq. (6.14).

6.5.3 Calculation of Eq. (6.17)

It is necessary to point out the properties of the complete first and second ellipse integrals K^* and E^*

$$K^*(k) \equiv \int_0^{\pi/2}\frac{d\phi}{\sqrt{1 - k^2\sin^2\phi}} = \int_0^{\pi/2}\frac{d\phi}{\sqrt{1 - k^2\cos^2\phi}} \tag{6.42}$$

$$E^*(k) \equiv \int_0^{\pi/2} \sqrt{1 - k^2 \sin^2 \phi} d\phi = \int_0^{\pi/2} \sqrt{1 - k^2 \cos^2 \phi} d\phi \qquad (6.43)$$

$$\frac{dK^*}{dk} = \frac{E^*}{kk'^2} - \frac{K^*}{k} \qquad (6.44)$$

$$\frac{dE^*}{dk} = \frac{E^* - K^*}{k}, \qquad (6.45)$$

where $k' = \sqrt{1 - k^2}$.

$$\int_0^{\pi/2} \frac{\cos^2 \theta^* \cos^2 \phi d\phi}{\sqrt{1 - \cos^2 \theta^* \sin^2 \phi}} = K^* - E^*. \qquad (6.46)$$

The free energy within an area of $x \times y = t \times L$ is expressed by

$$\begin{aligned}
F &= l\cos\theta \int_{t \times L} g_{LC} dA \\
&= Ll\cos\theta \int_0^t g_{LC} dx \\
&= 4Ll\cos\theta \int_0^{\pi/2} \left(\frac{k}{2} \sin^2 \theta \phi_x^2 - \frac{k_2}{2} \sin 2\theta \cos \phi \phi_x + \frac{k_2^2}{2k} \right) \frac{dx}{d\phi} d\phi, \qquad (6.47)
\end{aligned}$$

where

$$\frac{dx}{d\phi} = \frac{t}{4E^*} \sqrt{1 - \cos^2 \theta^* \cos^2 \phi} \quad \text{[see Eq. (6.14)]} \qquad (6.48)$$

$$\begin{aligned}
k_2 &= 2k_{20} \left(\chi - \frac{1}{2} \right) \quad [k_{20} = 2\pi k/p_0 \text{ and see Eq. (6.16)}] \\
&= \frac{2k \sin 2\theta \cos \phi E^*}{\eta t \sqrt{1 - \cos^2 \theta^* \cos^2 \phi}}. \qquad (6.49)
\end{aligned}$$

In the following Eq. (6.47) is calculated:

$$\begin{aligned}
\int_0^{\pi/2} \frac{k}{2} \sin^2 \theta \phi_x^2 \frac{dx}{d\phi} d\phi &= \frac{k \sin^2 \theta}{2} \int_0^{\pi/2} \frac{4E^*}{t\sqrt{1 - \cos^2 \theta^* \cos^2 \phi}} d\phi \\
&= \frac{2kE^* \sin^2 \theta}{t} \int_0^{\pi/2} \frac{d\phi}{\sqrt{1 - \cos^2 \theta^* \cos^2 \phi}} \quad \text{[see Eq. (6.48)]} \\
&= 2k \sin^2 \theta K^* E^*/t \quad \text{[see Eq. (6.19)]} \qquad (6.50)
\end{aligned}$$

$$\begin{aligned}
\int_0^{\pi/2} \frac{k_2}{2} \sin 2\theta \cos \phi \phi_x \frac{dx}{d\phi} d\phi &= \int_0^{\pi/2} \frac{k \sin^2 2\theta E^* \cos^2 \phi}{\eta t \sqrt{1 - \cos^2 \theta^* \cos^2 \phi}} d\phi \quad \text{[see Eq. (6.49)]} \\
&= \frac{4k \sin^2 \theta E^*}{t} \int_0^{\pi/2} \frac{\cos^2 \theta^* \cos^2 \phi}{\sqrt{1 - \cos^2 \theta^* \cos^2 \phi}} d\phi \\
&= \frac{4k \sin^2 \theta E^*}{t} (K^* - E^*) \qquad (6.51)
\end{aligned}$$

$$\int_0^{\pi/2} \frac{k_2^2 \, dx}{2k \, d\phi} d\phi = \int_0^{\pi/2} \frac{k \sin^2 2\theta \cos^2 \phi E^*}{2\eta^2 t \sqrt{1 - \cos^2 \theta^* \cos^2 \phi}} d\phi$$

$$= \frac{2k \sin^2 \theta E^*}{\eta t} (K^* - E^*). \tag{6.52}$$

Substituting Eqs. (6.50), (6.51) and (6.52) into Eq. (6.47) yields Eq. (6.17).

6.5.4 Calculation of Eq. (6.20)

Setting

$$f(\chi) = \chi \ln \chi + (1 - \chi) \ln(1 - \chi) + \alpha_0 \chi(1 - \chi), \tag{6.53}$$

the following set of differentials can be derived

$$f^{(1)}(\chi) = \ln \frac{\chi}{1 - \chi} + \alpha_0(1 - 2\chi)$$

$$f^{(2)}(\chi) = \frac{1}{\chi(1 - \chi)} - 2\alpha_0$$

$$f^{(3)}(\chi) = \frac{1}{(1 - \chi)^2} - \frac{1}{\chi^2}$$

$$f^{(4)}(\chi) = \frac{2}{\chi^3} + \frac{2}{(1 - \chi)^3}. \tag{6.54}$$

Using Eq. (6.54), one obtains the Taylor expansion of $f(\chi)$ at $\chi = 1/2$

$$f(\chi) = f\left(\frac{1}{2}\right) + f^{(1)}\left(\frac{1}{2}\right)\left(\chi - \frac{1}{2}\right) + \frac{1}{2!}f^{(2)}\left(\frac{1}{2}\right)\left(\chi - \frac{1}{2}\right)^2$$

$$+ \frac{1}{3!}f^{(3)}\left(\frac{1}{2}\right)\left(\chi - \frac{1}{2}\right)^3 + \frac{1}{4!}f^{(4)}\left(\frac{1}{2}\right)\left(\chi - \frac{1}{2}\right)^4$$

$$= \ln \frac{1}{2} + \frac{\alpha_0}{4} + (2 - \alpha_0)\left(\chi - \frac{1}{2}\right)^2 + \frac{4}{3}\left(\chi - \frac{1}{2}\right)^4 \tag{6.55}$$

Substituting Eq. (6.55) into Eq. (6.5), one gets Eq. (6.20).

6.5.4 Calculation of Eq. (6.21)

Eq. (6.20) can be rewritten in another form

$$F_{MIX} = \frac{L}{A_0} k_B T \int_0^{2\pi} \left[\ln \frac{1}{2} + \frac{\alpha_0}{4} + (2 - \alpha_0)\left(\chi - \frac{1}{2}\right)^2 + \frac{4}{3}\left(\chi - \frac{1}{2}\right)^4 \right] \frac{dx}{d\phi} d\phi, \tag{6.56}$$

where

$$\int_0^{2\pi} \left(\ln \frac{1}{2} + \frac{\alpha_0}{4} \right) \frac{dx}{d\phi} d\phi = \left(\ln \frac{1}{2} + \frac{\alpha_0}{4} \right) \int_0^t dx$$

$$= \left(\ln \frac{1}{2} + \frac{\alpha_0}{4} \right) t, \tag{6.57}$$

$$\int_0^{2\pi} (2 - \alpha_0) \left(\chi - \frac{1}{2} \right)^2 \frac{dx}{d\phi} d\phi = \int_0^{2\pi} (2 - \alpha_0) \frac{p_0^2 \sin^2 2\theta \, E^{*2} \cos^2 \phi}{4\pi^2 \eta^2 t^2 (1 - \cos^2 \theta^* \cos^2 \phi)} \cdot$$

$$\frac{t\sqrt{1 - \cos^2 \theta^* \cos^2 \phi}}{4E^*} d\phi$$

$$= (2 - \alpha_0) \left(\frac{p_0 \sin \theta}{\pi} \right)^2 \frac{E^*}{\eta t} \int_0^{\pi/2} \frac{\cos^2 \theta^* \cos^2 \phi}{\sqrt{1 - \cos^2 \theta^* \cos^2 \phi}} d\phi$$

$$= (2 - \alpha_0) \left(\frac{p_0 \sin \theta}{\pi} \right)^2 \frac{E^*}{\eta t} (K^* - E^*). \tag{6.58}$$

$$\int_0^{2\pi} \frac{4}{3} \left(\chi - \frac{1}{2} \right)^4 \frac{dx}{d\phi} d\phi = \frac{1}{3} \left(\frac{p_0 \sin \theta}{\pi t} \right)^4 \eta^{-2} \cos^4 \theta^* E^{*3} t \times$$

$$\int_0^{2\pi} \frac{\cos^4 \phi}{(1 - \cos^2 \theta^* \cos^2 \phi)^{3/2}} d\phi. \tag{6.59}$$

The calculation of the integral in Eq. (6.59) becomes the main task. Using Eqs. (6.42), (6.43), (6.44) and (6.45),

$$\frac{dE^*}{dk} = -k \int_0^{\pi/2} \frac{\sin^2 \phi}{\sqrt{1 - k^2 \sin^2 \phi}} d\phi \tag{6.60}$$

$$\frac{d^2 E^*}{dk^2} = -\int_0^{\pi/2} \frac{\sin^2 \phi}{\sqrt{1 - k^2 \sin^2 \phi}} d\phi - k^2 \int_0^{\pi/2} \frac{\sin^4 \phi}{(1 - k^2 \sin^2 \phi)^{3/2}} d\phi$$

$$= -\frac{E^* - K^*}{k^2} + \frac{1}{k} \left(\frac{dE^*}{dk} - \frac{K^*}{dk} \right). \tag{6.61}$$

But

$$\int_0^{\pi/2} \frac{\sin^2 \phi}{\sqrt{1 - k^2 \sin^2 \phi}} d\phi = \frac{K^* - E^*}{k^2}. \tag{6.62}$$

From the above equations, one obtains

$$k^2 \int_0^{\pi/2} \frac{\sin^4 \phi}{(1 - k^2 \sin^2 \phi)^{3/2}} d\phi = \frac{1}{k^2} \left[2(E^* - K^*) + \frac{k^2}{1 - k^2} E^* \right], \tag{6.63}$$

where $k = \cos \theta^*$. Thus

$$\int_0^{2\pi} \frac{\cos^4 \phi}{(1 - \cos^2 \theta^* \cos^2 \phi)^{3/2}} d\phi = 4 \int_0^{\pi/2} \frac{\sin^4 \phi}{(1 - k^2 \sin^2 \phi)^{3/2}} d\phi$$

$$= \cos^{-4} \theta^* (E^* - 2K^* + \sin^{-2} \theta^* E^*). \tag{6.64}$$

Substituting Eq. (6.64) into Eq. (6.59) yields

$$\int_0^{2\pi} \frac{4}{3} \left(\chi - \frac{1}{2} \right)^4 \frac{dx}{d\phi} d\phi = \frac{t\eta^{-2}}{3} (\sin^{-2} \theta^* E^{*4} - 2K^* E^{*3} + E^{*4}) \left(\frac{p_0 \sin \theta}{\pi t} \right)^4. \tag{6.65}$$

Substituting Eqs. (6.57), (6.58) and (6.65) into Eq. (6.56), and divided by the number of molecules $\dfrac{tL}{A}$, one obtains the Bragg-Williams mixing energy per molecule Eq. (6.21).

References

[1] A. Collet, M. J. Brienne, and J. Jacques, *Chem. Rev.*, **80** (1980) 215.

[2] L. Pasteur, *C. R. Acad. Sci. Paris*, **26** (1848) 535.

[3] M. E. Jungfleish, *J. Pharm. Chim.*, **5** (1882) 346.

[4] M. Lundquist, *Arkiv. Kem.*, **17** (1960) 183.

[5] M. Lundquist, *Arkiv. Kem.*, **21** (1962) 395.

[6] C. J. Eckhardt, N. M. Peachey, D. R. Swanson, J. M. Takacs, M. A. Khan, X. Gong, J. -H. Kim, J. Wang, and R. A. Uphau, *Nature*, **362** (1993) 614.

[7] J. Maclennan and M. Seul, *Phys. Rev. Lett.*, **69** (1992) 2082.

[8] X. Qiu, J. Ruiz-Garcia, K. J. Stine, C. M. Knobler, and J. V. Selinger, *Phys. Rev. Lett.*, **67** (1991) 703.

[9] J. V. Selinger, Z. G. Wang, R. F. Bruinsma, and C. M. Knobler, *Phys. Rev. Lett.*, **70** (1993) 1139.

[10] G. A. Hinshaw, R. G. Petschek, and R. A. Pelcovits, *Phys. Rev. Lett.*, **60** (1988) 1864.

[11] G. A. Hinshaw and R. A. Petschek, *Phys. Rev.*, **A39** (1989) 5914.

[12] Z. C. Ou-Yang and J. X. Liu, *Phys. Rev. Lett.*, **65** (1990) 1679.

[13] Z. C. Ou-Yang and J. X. Liu, *Phys. Rev.*, **A43** (1991) 6826.

[14] M. Iwamoto, C. X. Wu, and Z. C. Ou-Yang, *Phys. Rev.*, **E59** (1999) 586.

[15] P. G. de Gennes, *The Physics of Liquid Crystals*, Clarrendon Press, Oxford (1991).

[16] W. L. Bragg and E. J. Williams, *Proc. Roy. Soc.*, **A145** (1934) 699.

[17] R. Kubo, *Statistical Mechanics*, Kyoritsu, Tokyo (1970) p. 204 (in Japanese).

[18] W. Helfrisch, *Z. Naturforsch.*, **28a** (1973) 693.

[19] For a review, see, J. M. Schnur, *Science*, **262** (1993) 1669.

[20] M. A. Markowitz, J. M. Schnur, and A. Singh, *Chem. Phys. Lipids*, **62** (1992) 193.

[21] A. Sugimura, M. Iwamoto, and Z. C. Ou-Yang, *Phys. Rev.*, **E50** (1994) 614.

[22] M. Iwamoto, Y. Majima, H. Naruse, T. Noguchi, and H. Fuwa, *Nature*, **353** (1991) 645.

[23] I. Langmuir, *J. Chem. Phys.*, **1** (1933) 756.

[24] W. Zhao, C. X. Wu, and M. Iwamoto, *Phys. Rev.*, **E61** (2000) 6669.

[25] G. A. Korn and T. M. Korn, *Mathematical Handbook for Scientists and Engineers*, Mcgraw-Hill, New York (1961) p. 322.

CHAPTER 7

NONLINEAR EFFECTS

In recent years, a variety of experimental methods including scattering, spectroscopic, and electrical techniques have been developed to study the molecular structure and orientational phase transition of insoluble monolayers at the air-water interface. [1, 2, 3, 4, 5] However, there has been a lack of detailed information at the molecular level, especially, the symmetries of the molecules. On the other hand, the second-harmonic generation (SHG) from a monolayer at air-water surface possesses the obvious advantage of being intrinsically monolayer-specific as it is forbidden in centrosymmetric bulk water and air. [6, 7] Therefore, SHG, sum-, and difference-frequency generations (SFG and DFG) from a monolayer absorbed on liquid surface have been extensively and successfully investigated. [8, 9, 10, 11, 12] The SFG intensity generated from such a monolayer is governed by a surface nonlinear polarization $\overrightarrow{P}^{N}$ in the general form

$$P_i^N(\omega_1 + \omega_2) = \chi_{ijk}(-(\omega_1 + \omega_2), \omega_1, \omega_2) E_j(\omega_1) F_k(\omega_2), \tag{7.1}$$

where $\overrightarrow{E}(\omega_1)$ and $\overrightarrow{F}(\omega_2)$ are the fundamental optical fields at frequency ω_1 and ω_2, respectively. The SHG and DFG are, respectively, the cases of $\omega_1 = \omega_2$ and $\omega_2 \to -\omega_2$.

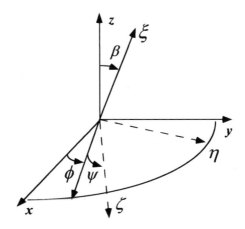

Figure 7.1 Configuration of the two coordinate systems and Euler angles.

The second-order susceptibility (SOS) $\chi^{(2)} \equiv [\chi_{ijk}]$ is related to the molecular SOS tensor $\alpha^{(2)} \equiv [\alpha_{\lambda\mu\nu}]$ by

$$\chi_{ijk} = N_s \langle T^{\lambda\mu\nu}_{ijk} \rangle \alpha_{\lambda\mu\nu} \qquad (7.2)$$

where N_s is the surface density of the molecules, $T^{\lambda\mu\nu}_{ijk}$ describes the coordinate transformation between the molecular (ζ, η, ξ) system and the lab (x, y, z) system, and $\langle\ \rangle$ denotes a thermodynamics average over the molecular orientations. Figure 7.1 shows the configuration of the two coordinate systems and the Euler angles. For the expression Eq. (7.1), for simplicity no account is taken of the local field correction facotors such as Lorentz facotor which arise due to the surface, but also, significantly, due to dipole-dipole interaction with neighbouring monolayer molecules, where the latter will be obviously concentration dependent. This simplification does not lose physics underlying here because one way of local field correction is introducing a factor to the right-hand side of Eq. (7.1), which will not influence the present analysis and the following discussion. [7] From the SHG measurements, one can deduce $\chi^{(2)}$, but to obtain $\alpha^{(2)}$ one has to have the knowledge of $\langle T^{\lambda\mu\nu}_{ijk} \rangle$ which has been so far a task in mathematics. Due to $T^{\lambda\mu\nu}_{ijk} = R^\lambda_i R^\mu_j R^\nu_k$ with $R(\phi, \beta, \psi) = [R^\lambda_i]$ being the usual Euler rotation matrix

$$R^\lambda_i = \begin{pmatrix} \cos\phi\cos\psi - \sin\phi\cos\beta\sin\psi, & -\cos\phi\sin\psi - \sin\phi\cos\beta\cos\psi, & \sin\phi\sin\beta \\ \sin\phi\cos\psi + \cos\phi\cos\beta\sin\psi, & -\sin\phi\sin\psi + \cos\phi\cos\beta\cos\psi, & -\cos\phi\sin\beta \\ \sin\beta\sin\psi, & \sin\beta\cos\psi, & \cos\beta \end{pmatrix}$$

$$(7.3)$$

between (x, y, z) and (ζ, η, ξ) with Euler angles (ϕ, β, ψ) defined as the convention, [13] one has to perform a lengthy average calculation for $\langle T \rangle$ with 9^3 components with a distribution $f(\phi, \beta, \psi)$. Unfortunately, such calculation has not been done in general, but only in a special case where $\alpha^{(2)}$ is dominated by a single component $\alpha_{\xi\xi\xi}$ along a molecular axis $\boldsymbol{\xi}$. [8] The independent non-vanishing elements of $\chi^{(2)}$ then reduce to two. [8] Obviously, such a simplification can deal only with a $C_{\infty v}$ symmetric monolayer, a nonchiral uniaxial surface.

Very recently, the SHG circular-dichroism (CD) has been found from a monolayer composed of oriented chiral molecules of R- or S-2, 2'-dihydroxyl-1, 1'binaphthyl (R- or S-BN) [14] and has been explained theoretically by the electric dipole-allowed $\chi^{(2)}$ terms for an isotropic surface invariant with rotations about the perpendicular z axis, [15] i.e., the 27 elements of the tensor $\chi^{(2)}$ can be reduced to four nonzero and independent elements; [16]

$$\begin{aligned} \chi_1 &= \chi_{zzz} \\ \chi_2 &= \chi_{zxx} = \chi_{zyy} \\ \chi_3 &= \chi_{xzx} = \chi_{xxz} = \chi_{yzy} = \chi_{yyz} \\ \chi_4 &= \chi_{xzy} = \chi_{xyz} = -\chi_{yzx} = -\chi_{yxz}, \end{aligned} \qquad (7.4)$$

where χ_4 characterizes the chirality of the monolayer. The relations given in Eq. (7.4) is derived from the assumption that the input photons are degenerated (i.e., $\alpha_{\lambda\mu\nu} =$

$\alpha_{\lambda\nu\mu}$). [16] From Eq. (7.1), such an assumption is not completely used for SFG and DFG, both of which give rise at least to an additional independent element, since $\chi_{xzx} = \chi_{yzy} \neq \chi_{xxz} = \chi_{yyz}$. In other words, Eq. (7.4) is still not a general case for a uniaxial surface and, moreover, not related to the molecular orientation. The latter is remarkably lacking in physics. For example, the chiral term χ_4 is believed not related to polar orientational order by imaging a helix having no difference between its up and down geometries. Thus, the general question can be posed as follows: What is the general expression of $\chi^{(2)}$ associated with $\alpha^{(2)}$ molecular orientation for a monolayer at the liquid-air interface, and can it be used to distinguish chiral and achiral monolayers and then to make a spectroscopy to measure the molecular SOS $\alpha^{(2)}$?

In this chapter, the complete expression for the macroscopic SOS $\chi^{(2)}$ of a monolayer with C_∞ symmetry on a material surface as functions of the molecular SOS tensor $\alpha^{(2)}$ and the orientational order parameters $S_n = \langle P_n(\cos\theta) \rangle$ will be shown. [17] Here θ is the tilt angle of the molecular main axis $\boldsymbol{\xi}$ from the normal $\boldsymbol{z}$ of the monolayer, and P_n is the n-th Legendre polynomial. With this definition, the chiral (nonchiral) terms of $\chi^{(2)}$ are clearly distinguished by their association with S_0, S_2 (S_1 and S_3) and the SHG-CD experiment reported by Byers *et al.* [14, 15] may be very well understood. It has been shown that the uniaxially orientational order parameters can be calculated or measured as functions of the molecular area A. [17] Therefore, upon the variation of the $S_1(A)$, $S_2(A)$ and $S_3(A)$ by compressing the molecular area A, the SHG spectroscopy from the monolayer allows the study of the molecular SOS tensor $\alpha^{(2)}$ and its structural symmetry. With this hypothesis a SHG experiment of a 5CB (4-cyno-4'-5-alkyl-biphenyl) monolayer on an air-water interface will be discussed.

7.1 SOS in Orientational Order Parameters for C_∞ Monolayers

Phenomenologically, the general expression for $\chi^{(2)}$ of C_∞-symmetry has been given by Giordmaine. [18] By changing Eq. (7.1) from tensor notation to matrix notation, [19]

$$P_i^N = s_{ij}f_j + a_{ij}\overline{f}_j \qquad (i = 1 - 3, \quad j = 1 - 6) \tag{7.5}$$

with

$$
\begin{aligned}
s_{i1} &= \chi_{i11} \\
s_{i2} &= \chi_{i22} \\
s_{i3} &= \chi_{i33} \\
s_{i4} &= \chi_{i23} + \chi_{i32} \\
s_{i5} &= \chi_{i31} + \chi_{i13} \\
s_{i6} &= \chi_{i12} + \chi_{i21} \\
a_{ij} &= 0 \quad for\ j = 1 - 3 \\
a_{i4} &= \chi_{i23} - \chi_{i32}
\end{aligned}
$$

$$a_{i5} = \chi_{i31} - \chi_{i13}$$
$$a_{i6} = \chi_{i12} - \chi_{i21}$$
$$f = \frac{1}{2}(2E_1F_1, 2E_2F_2, 2E_3F_3, E_2F_3 + E_3F_2, E_3F_1 + E_1F_3, E_1F_2 + E_2F_1)$$
$$\overline{f} = \frac{1}{2}(0, 0, 0, E_2F_3 - E_3F_2, E_3F_1 - E_1F_3, E_1F_2 - E_2F_1).$$

Here $\chi^{(2)}$ of C_∞ symmetry is given by the following two matrices [18]

$$s = \begin{pmatrix} 0 & 0 & 0 & s_{14} & s_{15} & 0 \\ 0 & 0 & 0 & s_{15} & -s_{14} & 0 \\ s_{31} & s_{31} & s_{33} & 0 & 0 & 0 \end{pmatrix} \qquad a = \begin{pmatrix} 0 & 0 & 0 & a_{14} & a_{15} & 0 \\ 0 & 0 & 0 & -a_{15} & a_{14} & 0 \\ 0 & 0 & 0 & 0 & 0 & a_{36} \end{pmatrix}, \quad (7.6)$$

with 7 independent nonzero elements. Hereafter suffixes (1,2,3) will be referred to (x, y, z) in lab system and (ζ, η, ξ) in molecular system. It shall be proved that for a uniaxial monolayer, the definition of $\chi^{(2)}$ given in Eq. (7.2) does lead to Eq. (7.6) without any additional assumption on $\alpha^{(2)}$ because in the electric-dipole approximation $\alpha^{(2)}$ concerns only with the electronic structure of the molecule and should not be supposed to have some symmetry. On the other hand, the uniaxial property of the monolayer at the air-liquid interface is related to the repulsive interaction between molecules of the monolayer and the interaction between the dipolar molecules and the interfaces. [17] In other words, the average of $\langle T_{ijk}^{\lambda\mu\nu} \rangle$ in Eq. (7.2) for a uniaxial surface is written as

$$\int T_{ijk}^{\lambda\mu\nu}(\phi, \beta, \psi) f(\beta) \sin \beta d\phi d\beta d\psi / \int f(\beta) \sin \beta d\phi d\beta d\psi, \qquad (7.7)$$

where $T_{ijk}^{\lambda\mu\nu}(\phi, \beta, \psi)$ has been defined in the beginning and the orientation distribution function is independent of ϕ and ψ specifically in the case of C_∞ symmetry, otherwise, C_∞ symmetry can be broken (e.g. $f(\beta, \psi)$ must result in biaxial ordering [20]). Here angle β is just θ, the tilt angle of the molecular axis ξ from surface normal z. [13]

The generalized method for calculating Eq. (7.5) has been presented by Andrews et. al, [21, 22] and the result for monolayers at the water surface has been reported, assuming the phase and the Boltzmann-weighted rotational averages. However, as the main interest here is to show explicitly the general physics of $\chi^{(2)}$ associated with the molecular SOS tensor $\alpha^{(2)}$ for chiral and achiral monolayers, which has not yet been discussed before, it is important to reconstruct $\chi^{(2)}$. And without neglecting physics underlying here, it shall be shown that Eq. (7.7) leads exactly to the form as Eq. (7.6), if neglecting the intermolecular interaction, i.e., $f(\beta) = 1$. More specifically, the seven independent nonzero elements in Eq. (7.7) can be expressed as the following important relations: [23]

$$s_{14} = \frac{N_s}{2} S_2 (\sigma_{14} - \sigma_{25}),$$
$$s_{15} = \frac{N_s}{5}(S_1 - S_3)(2\sigma_{33} - \sigma_{32} - \sigma_{31}) + \frac{N_s}{10}(3S_1 + 2S_3)(\sigma_{24} + \sigma_{15}),$$

$$s_{31} = \frac{N_s}{10}(S_1 - S_3)(2\sigma_{33} - \sigma_{24} - \sigma_{15}) + \frac{N_s}{10}(4S_1 + S_3)(\sigma_{32} + \sigma_{31}),$$

$$s_{33} = \frac{N_s}{5}(S_1 - S_3)(\sigma_{32} + \sigma_{31} + \sigma_{24} + \sigma_{15}) + \frac{N_s}{5}(3S_1 + 2S_3)\sigma_{33},$$

$$a_{14} = \frac{N_s}{3}(\lambda_{14} + \lambda_{25} + \lambda_{36}) + \frac{N_s}{6}S_2(\lambda_{14} + \lambda_{25} - 2\lambda_{36}),$$

$$a_{15} = \frac{N_s}{2}S_1(\lambda_{15} - \lambda_{24}),$$

$$a_{36} = \frac{N_s}{3}(\lambda_{14} + \lambda_{25} + \lambda_{36}) + \frac{N_s}{3}S_2(2\lambda_{36} - \lambda_{14} - \lambda_{25}), \tag{7.8}$$

where the two 3×6 matrices, (σ_{ij}) and (λ_{ij}), are defined from the molecular SOS tensor $\alpha^{(2)}$ in the same way as conventional contracted notation of (s_{ij}) and (a_{ij}) defined from $\chi^{(2)}$, i.e., $\sigma_{11} = \alpha_{111}$, $\sigma_{14} = \alpha_{123} + \alpha_{132}$, $\lambda_{14} = \alpha_{123} - \alpha_{132}$ etc. Here S_1, S_2, and S_3 are the molecular order parameters as defined above.

7.2 Chirality Representation

As the first application of the obtained general expression Eq. (7.8), the specific properties of chiral and achiral monolayers and their association with molecular chirality will be discussed. From the association with the order parameters, the seven terms given in Eq. (7.8) can be distinguished by two groups, the chiral terms (s_{14}, a_{14}, a_{36}) and the nonchiral terms $(s_{15}, s_{31}, s_{33}, a_{15})$. The reason has been roughly pointed out in the beginning on the feature of a helix, but Eq. (7.8) can provide more intrinsic mechanism from molecular level. Using tensor notation, it is found that $s_{14} = (N_s/2)S_2(\alpha_{123} + \alpha_{132} - \alpha_{213} - \alpha_{231})$, in which all terms of α_{ijk} are obviously chiral, i.e. $(\alpha_{123}, \alpha_{132}, \alpha_{213}, \alpha_{231})$ changes to $(-\alpha_{123}, -\alpha_{132}, -\alpha_{213}, -\alpha_{231})$ whenever the system $(\hat{1}, \hat{2}, \hat{3})$ changes to $(-\hat{1}, \hat{2}, \hat{3})$. On the other hand, it is found that $s_{15} = (N_s/5)(S_1 - S_3)(2\alpha_{333} - \alpha_{322} - \alpha_{311}) + (N_s/10)(3S_1 + 2S_3)(\alpha_{113} + \alpha_{223} + \alpha_{131} + \alpha_{232})$, in which all α_{ijk} are molecularly nonchiral, i.e., the components α_{333}, α_{322}, etc., are unchanged under the above operation for system $(\hat{1}, \hat{2}, \hat{3})$. Given these the symmetry of $\chi^{(2)}$ by the molecular structures, i.e. $\alpha^{(2)}$, can now be discussed. If molecules are nonchiral (i.e. $s_{14} = a_{14} = a_{36} = 0$), one can find that the two matrices given in Eq. (7.6) reduce precisely to those for the case of $C_{\infty v}$ symmetry (see the complete tabulations by Giordmaine [18]) as what it should be. Equation (7.8) can also serve to discuss the symmetry of $\chi^{(2)}$ with molecular orientation. For instance, for nonpolar molecules the absorption on the interface is regardless of head or tail, which means $S_1 = S_3 = 0$, and hence the nonchiral terms in Eq. (7.8) are vanishing (i.e., $s_{15} = s_{31} = s_{33} = a_{15} = 0$). In this case, the textures of Eq. (7.6) reduce to D_{∞} symmetry as shown in the mentioned tabulations. The D_{∞} symmetry is that for a film of cholesteric liquid crystals. From Eq. (7.8), one can also find that monolayers consisting of chiral molecules always possess optical nonlinearity even when randomly distributed (i.e., $S_1 = S_2 = S_3 = 0$). In this case Eq. (7.6) leads to the optically active isotropic case (see $[\infty, \infty]$ tabulation, *ibid.*).

7.3 SHG-CD Effect

Now consider the SHG-CD effect of a chiral monolayer reported by Petralli-Mallow and Byers. [14, 15] A schematic diagram is given by Fig. 7.2. Although the general expression Eq. (7.8) cannot result precisely in Eq. (7.4) without the assumption of $\alpha_{ijk} = \alpha_{ikj}$ (e.g., Equation (7.8) gives no relations of $\chi_{xxz} = \chi_{xzx}$ etc. by Kleinman's rule [24]), by viewing the air-quartz interface as a C_∞-symmetric interface and with the similar derivation process as reported by Byers *et al.*, [15] using Eq. (7.8) with the same appearance of SHG-CD effect, e.g. with the same geometry and symbolism given by Byers *et al.*, the intensity for s-polarized component of the reflected SHG wave from circularly polarized incident light can be obtained as

$$I_{2\omega}^s = \frac{8\pi^3\omega^2}{c^3} \tan^2\theta_i \mid \pm i s_{15} - s_{14}\cos\theta_i \mid^2 I_\omega^2, \tag{7.9}$$

where upper (lower) sign represents right (left) circularly polarized incident light. From the definition of SHG-CD, $I_{SHG-CD} = 2(I_{left}^{SHG} - I_{right}^{SHG})/(I_{left}^{SHG} + I_{right}^{SHG})$, where suffixes left and right represent the handedness of circularly polarized incident light, [15] Eqs. (7.8), and (7.9), two effects can be verified immediately: (1) For the two molecular enantiomers of left- and right-handed chirality (denoted as L- and R-enantiomers), I_{SHG-CD} (L-enantiomer)$=-I_{SHG-CD}$ (R-enantiomer) due to s_{14} (R-enantiomer)$=-s_{14}$ (L-enantiomer) and s_{15} (R-enantiomer)$=s_{15}$ (L-enantiomer). The latter two relations can be seen from their associations with $\alpha^{(2)}$, i.e., s_{14} being chiral and s_{15} achiral. (2) For a monolayer of the same handed enantiomer, I_{SHG-CD} must change sign when the light incident from up-side changes to that from down-side. The reason is apparent from Eq. (7.8) that the changes in two geometries are equivalent to the changes of the orientation, $S_1, S_3 \rightarrow -S_1, -S_3$ but S_2 is still unchanged, and thence $s_{14}^{up} = s_{14}^{down}$ and $s_{15}^{up} = -s_{15}^{down}$, which leads to $I_{SHG-CD}^{up} = -I_{SHG-CD}^{down}$. Both effects have been verified by the experimental results by Byers *et al.* [15] Although some explanations were given by them, the current one contains more physics in association with chirality and orientational orders for such kind of monolayers.

7.4 SHG-MDC Measuring System

The advantage of the general expression shown in Eq. (7.8) also leads to a great potential use to determine molecular SOS $\alpha^{(2)}$ from the SHG spectroscopy from a monolayer at air-water interface by compressing the molecular area of the monolayer. With the experimental setup given by Fig. 7.3 and the geometry of a C_∞-symmetric monolayer at the air-water surface shown in the lower inset of Fig. 7.4, in our previous work [17] molecular orientational orders have been investigated based on the repulsive interaction between monolayer-molecules (represented by $0 \le \theta = \beta \le \theta_A \equiv \arcsin(\sqrt{A/A_0})$,

Chiral R-2,2'-dihydroxy-1,1'binaphythyl (BN)

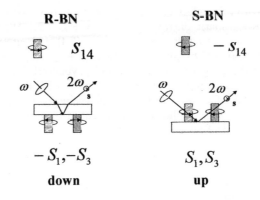

Figure 7.2 Schematic diagram for SHG-CD effect of BN molecules as observed by Byers *et al.* (Ref. [15]).

where $A_0 = \pi l^2$ and l is the partial length of the molecules along the long axis above the water surface) and the interaction between dipolar molecules and water (represented by the energy $W(\theta) = -(\mu^2/16\pi\epsilon_0\epsilon_m l^3 \cos\theta)[(\epsilon_w - \epsilon_m)/(\epsilon_w + \epsilon_m)]$ where μ is the molecular dipole moment, ϵ_0, ϵ_m and ϵ_w are the permittivities of free space, monolayer and water, respectively). With the apparent distribution function $f(\theta) = \exp[-W(\theta)/k_BT]$, k_B being the Boltzmann's constant, and T the absolute temperature, any-order orientational order parameter can be deduced by the integration

$$S_n = \int_0^{\theta(A)} P_n(\cos\theta)f(\theta)\sin\theta d\theta \Big/ \int_0^{\theta(A)} f(\theta)\sin\theta d\theta. \qquad (7.10)$$

It is shown that S_n are mainly attributed to the repulsive interaction and weakly depend on the dipolar interaction (see Figs. 1 and 2 in Ref.[16]). Therefore, as a first approximation, let $\mu \to 0$ and have from Eq. (7.10) the analytic results of

$$S_1 = \frac{1}{2}(1 + \cos\theta_A)$$

$$S_2 = \frac{1}{2}\cos\theta_A(1 + \cos\theta_A)$$

$$S_3 = \frac{1}{8}(5\cos^2\theta_A - 1)(1 + \cos\theta_A).$$

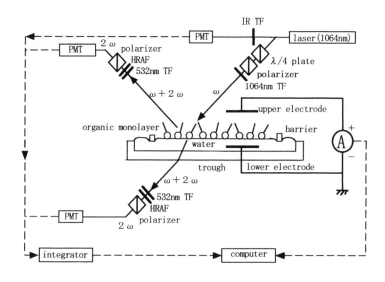

HRAF: heat rays absorption filter
PMT: photomultiplier
TF: transmittance filter

Figure 7.3 Experimental setup of the SHG-MDC simultaneous meansurement.

Substituting the above equations to Eq. (7.8) leads $\chi^{(2)}$ to an apparent function of A and $\alpha^{(2)}$ only. Therefore, by a best fitting on the variation curve of the SHG intensity with molecular area A, the molecular SOS $\alpha^{(2)}$ can be determined. As an example, in what follows, an SHG experiment performed by the authors' group shall be analyzed. As shown in Fig. 7.2, the monolayers were prepared by spreading achiral 5CB molecules on the surface of water in a Langmuir trough. With a linear polarized light (Nd: YAG laser from BIG SKY LASER TECH. with $\lambda = 1.064\ \mu$m, maximum-50 mJ pulses of half width 7 nsec and 10 Hz) incident from the air side onto the monolayer, both the reflected and transmitted SHGs from the monolayer were observed. The upper inset in Fig. 7.3 presents a reflected SHG intensity spectrum of $pp \rightarrow p$ polarized generation by compressing molecular area from $A = A_0 = 90$ Å^2 [17] to 36 Å^2. To use the above theoretical prediction to analyze the spectrum, we first use Eq. (7.8) to rewrite Eq. (7.1) as a vectorial form of $\overrightarrow{P}^N = \overrightarrow{P}^N_{ch} + \overrightarrow{P}^N_{ach}$ with $\overrightarrow{P}^N_{ch}$ associated to the chirality of the monolayer, and $\overrightarrow{P}^N_{ach}$ to the nonchirality:

$$
\overrightarrow{P}^N_{ch} = \frac{1}{2}s_{14}[(\overrightarrow{E}\cdot\overrightarrow{n})(\overrightarrow{F}\times\overrightarrow{n}) + (\overrightarrow{F}\cdot\overrightarrow{n})(\overrightarrow{E}\times\overrightarrow{n})] + \frac{1}{2}a_{14}(\overrightarrow{E}\times\overrightarrow{F})
$$
$$
+\frac{1}{2}(a_{36} - a_{14})\overrightarrow{n}\cdot(\overrightarrow{E}\times\overrightarrow{F})\overrightarrow{n}, \tag{7.11}
$$

$$
\overrightarrow{P}^N_{ach} = (s_{33} - s_{15} - s_{31})(\overrightarrow{n}\cdot\overrightarrow{E})(\overrightarrow{n}\cdot\overrightarrow{F})\overrightarrow{n} + s_{31}(\overrightarrow{E}\cdot\overrightarrow{F})\overrightarrow{n}
$$
$$
+\frac{1}{2}s_{15}[(\overrightarrow{n}\cdot\overrightarrow{F})\overrightarrow{E} + (\overrightarrow{n}\cdot\overrightarrow{E})\overrightarrow{F}] + \frac{1}{2}a_{15}(\overrightarrow{E}\times\overrightarrow{F})\times\overrightarrow{n}, \tag{7.12}
$$

where $\overrightarrow{n}$ is the unit normal of the monolayer (i.e. $\overrightarrow{n} = z$). From Eq. (7.11), one can see that s_{14}, a_{14}, and a_{36} must be pseudo-scalar because all vectors of $\overrightarrow{P}^N$, $\overrightarrow{E}$, $\overrightarrow{F}$ and $\overrightarrow{n}$ are polar vectors. For SHG from a nonchiral monolayer such as 5CB monolayer, by letting $s_{14} = 0$ and $\overrightarrow{F} = \overrightarrow{E}$, Eq. (7.11) and Eq. (7.12) reduce to the following simple form

$$
\overrightarrow{P}^N = (\overline{A} - B)(\overrightarrow{n}\cdot\overrightarrow{E})^2\overrightarrow{n} + B(\overrightarrow{n}\cdot\overrightarrow{E})\overrightarrow{E} + C(\overrightarrow{E}\cdot\overrightarrow{E})\overrightarrow{n} \tag{7.13}
$$

with $\overline{A} = s_{33} - s_{31}$, $B = s_{15}$, and $C = s_{31}$. Eq. (7.13) is identical to the phenomenological result given by Wang [25] specially for the surface of an isotropic medium. This not only gives evidence again for Eq. (7.8) but also provides a fortunate chance for the current task: For the 5CB SHG experiment, we do not need to calculate afresh. The theoretical result for the optical electric field of $pp \rightarrow p$ (input/output) polarization combination is a special case of the rigorous results given by Wang, [25] which is the observation in the experiment,

$$
E^p(2\omega) = \frac{32\pi\omega}{c}\frac{E_0^2\sin\theta_i\cos^2\theta_i}{\sqrt{\epsilon_s}g_{OS}^2g_{OT}}[\overline{A}\sin^2\theta_i\sqrt{\epsilon_T}/\sqrt{\epsilon_S} + C\sqrt{\epsilon_S\epsilon_T} - B\cos\theta_S\cos\theta_T], \tag{7.14}
$$

where θ_i is the angle of incident, θ_S and θ_T are those of the reflected and transmitted second-harmonic waves (i.e. $\sin\theta_i = \sqrt{\epsilon_S}\sin\theta_S = \sqrt{\epsilon_T}\sin\theta_T$), respectively, and the Fresnel corrections are $g_{OS} = \sqrt{\epsilon_S}\cos\theta_i + \cos\theta_S$ and $g_{OT} = \sqrt{\epsilon_T}\cos\theta_i + \cos\theta_T$. From the measured $I_{SHG}^p \propto | E^p(2\omega) |^2$, we can determine A, B, and C (i.e., s_{15}, s_{33}, and s_{31}). Here, as an instructive example, we only use it to check the one-component model for $\alpha^{(2)}$ presented by Heinz. [8] By assuming $\alpha^{(2)}$ dominated by the single component α_{333}, from Eq. (7.8) and the relations of $S_1 - S_3$ with molecular area A mentioned above, we have

$$
\overline{A} = \frac{N_s}{8}(1 + x)(1 + 3x^2)\alpha_{333}
$$

$$B = \frac{N_s}{4}(1 - x^2)(1 + x)\alpha_{333}$$

$$C = \frac{B}{2},$$

where $x^2 \equiv \cos^2\theta_A = 1 - A/A_0$ and N_s ($= 1/A$) is the surface density of molecules. Substituting the above results into Eq. (7.14), and letting $\sqrt{\epsilon_S} \approx 1$ (at air), $\sqrt{\epsilon_T} = 1.33$ (at water), $\theta_i = 60°$ (from experiment by the authors' group), and $A_0 = 90$ Å^2, we obtain the SHG-A spectrum of 5CB monolayer at air-water interface

$$\frac{I_{SHG}^p(2\omega, A/A_0)}{I_{SHG}^p(2\omega, 0.4)} = \left[\frac{2.54 - 1.54(A/A_0)}{1 - \sqrt{1 - A/A_0}}\right]^2 \left(\frac{1 - \sqrt{0.6}}{1.924}\right)^2. \qquad (7.15)$$

The numerical result shows good agreement with the experimental result, as shown in

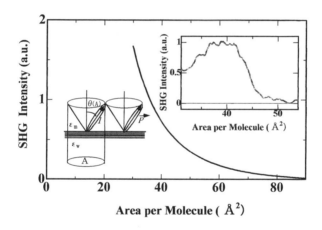

Figure 7.4 The calculated SHG intensity as a function of area per molecule for 5CB monolayer at the air-water interface. The polarization combination is chosen to 60°-in/p-out. The upper inset shows the experimental result and lower is the geometry of the orientation for the monolayer.

Fig. 7.4. This confirms the assumption in the early work [8] on one component model for $\alpha^{(2)}$ in the 5CB case. In the above calculation the minimum molecular area $A_{min} = 36$ Å^2 with $A_{min}/A_0 = 0.4$ for the 5CB monolayer is obtained by experiment. When A

is reduced below A_{min}, the monolayer is destroyed as composed of a monolayer plus some interdigitated bilayers. [26] The bilayers possess centrosymmetry, therefore, SHG arising from the multilayers decreases and as demonstrated in the upper inset of Fig. 7.4.

7.5 Quantum Mechanical Analysis of Photoisomerization

In this section, the photoizomerization of polyamic acid salts containing azobenzene derivatives (AZBPAA) monolayers formed on the water surface, which have been successfully detected by MDC measurement as shown in Fig. 3.16, will be explained. [27] $Trans{\rightarrow}cis$ and $cis{\rightarrow}trans$ photoinduced isomerizations were observed by imposing ultraviolet light centered at 380 nm (λ_1) and visible light centered at 450 nm (λ_2), respectively. Such phenomena are strongly related to the stereostructure of molecules in monolayers. For a monolayer system $\hat{H}_0$ in circularly polarized light, the Hamiltonian $\hat{H}^{int}(t)$ is written as [6]

$$\hat{H} = \hat{H}_0 + \hat{H}^{int}(t) + \hat{H}_{random}. \tag{7.16}$$

In the semiclassical approach, H_0 is the Hamiltonian of the unperturbed material system with eingenstates $\mid n\rangle$ and eingenenergies E_n so that $H_0 \mid n\rangle = E_n \mid n\rangle$, H^{int} is the interaction Hamiltonian describing the interaction of light with monolayer films, which depends on the handedness of the circularly polarized light, and H_{random} is a Hamiltonian describing the random perturbation on the system by the thermal reservoir around the system, which is responsible for the relaxations of material excitations. For simplicity, only the right handed circularly polarized light is considered. The perturbation energy as a result of the interaction between the dipole moment and the external light is given by

$$\begin{aligned} \hat{H}^{int} &= e\boldsymbol{r} \cdot \boldsymbol{E}^+ = -E(\mu_x \cos\omega t + \mu_y \sin\omega t) \\ &= \hat{H}^{int}(\omega)e^{-i\omega t} + \text{c.c.}, \end{aligned} \tag{7.17}$$

where c.c. refers to the complex conjugate of the first term and

$$\hat{H}^{int}(\omega) = -\frac{E}{2}(\mu_x + i\mu_y)$$

In the Heisenberg representation, the equation of motion for $\boldsymbol{\mu}$ can be readily obtained as:

$$\frac{\partial\boldsymbol{\mu}}{\partial t} = \frac{1}{i\hbar}[\hat{H}, \boldsymbol{\mu}]. \tag{7.18}$$

The Hamiltonian $\hat{H}_{random}$ in Eq. (7.16) is responsible for the relaxation of the perturbed dipole moment $\boldsymbol{\mu}$ back to thermal equilibrium. The relaxation process can also be depicted by

$$\left(\frac{\partial\boldsymbol{\mu}}{\partial t}\right)_{relax} = \frac{1}{i\hbar}[\hat{H}_{random}, \boldsymbol{\mu}]. \tag{7.19}$$

The relaxation is a result of the interaction with the thermal reservoir. Here the transition is expected to obey an exponential decay

$$\left(\frac{\partial \mu_{nn'}}{\partial t}\right)_{relax} = -\frac{\mu_{nn'}}{(\tau)_{nn'}}, \tag{7.20}$$

where τ is called the transverse relaxation time. To find the dipole moment of the perturbed system, the dipole moment is expanded into various orders:

$$\boldsymbol{\mu} = \boldsymbol{\mu}^{(0)} + \boldsymbol{\mu}^{(1)} + \boldsymbol{\mu}^{(2)} + \cdots, \tag{7.21}$$

where $\boldsymbol{\mu}^{(0)}$ is the permanent dipole moment with $\boldsymbol{\mu}^{(0)} = \boldsymbol{\mu}_0$, and $\boldsymbol{\mu}^{(1)}$ and $\boldsymbol{\mu}^{(2)}$ are respectively the additional first-order effect and second order effect of the dipole moment, induced by the perturbation $\hat{H}^{int}$. Substituting Eq. (7.16), Eq. (7.19), and Eq. (7.21) into Eq. (7.18),

$$\frac{\partial \boldsymbol{\mu}^{(1)}}{\partial t} = \frac{1}{i\hbar}[(\hat{H}_0, \boldsymbol{\mu}^{(1)}) + (\hat{H}^{int}, \boldsymbol{\mu}^{(0)})] + \left(\frac{\partial \boldsymbol{\mu}^{(1)}}{\partial t}\right)_{relax}, \tag{7.22}$$

$$\frac{\partial \boldsymbol{\mu}^{(2)}}{\partial t} = \frac{1}{i\hbar}[(\hat{H}_0, \boldsymbol{\mu}^{(2)}) + (\hat{H}^{int}, \boldsymbol{\mu}^{(1)})] + \left(\frac{\partial \boldsymbol{\mu}^{(2)}}{\partial t}\right)_{relax}, \tag{7.23}$$

and so on. Using the relationships

$$\left(\frac{\partial \boldsymbol{\mu}^{(1)}(\omega)}{\partial t}\right)_{nn'} = -i\omega \mu_{nn'}^{(1)}(\omega)$$

$$[\hat{H}_0, \boldsymbol{\mu}^{(1)}]_{nn'} = \hbar \omega_{nn'} \mu_{nn'}^{(1)}$$

$$[\hat{H}^{int}(\omega), \boldsymbol{\mu}^{(0)}]_{nn'} = H^{int}(\omega)_{nn'}(\mu_{n'n'}^{(0)} - \mu_{nn}^{(0)}), \tag{7.24}$$

and comparing the terms with the same time-dependent parts in Eqs. (7.22) and (7.23), the first-order transition dipole moments for ω from $|n\rangle$ to $|n'\rangle$ is explicitly given by

$$\mu_{nn'}^{(1)}(\omega) = \frac{H^{int}(\omega)_{nn'}(\mu_{n'n'}^{(0)} - \mu_{nn}^{(0)})}{\hbar(\omega - \omega_{nn'} + i/(\tau)_{nn'})}. \tag{7.25}$$

The first-order Maxwell displacement current I is produced by the direct transition from state $|n\rangle$ to state $|n'\rangle$, which corresponds to the differentiation of the transition moment with respect to time: [28]

$$\begin{aligned}
I^{(1)}(t) &\propto \frac{d\boldsymbol{\mu}_{nn'}(t)}{dt} \\
&= \frac{d}{dt}[\boldsymbol{\mu}_{nn'}^{(1)}(\omega)e^{-i\omega t} + \text{c.c.}] \\
&= \frac{-i\omega H^{int}(\omega)_{nn'}(\mu_{n'n'}^{(0)} - \mu_{nn}^{(0)})}{\hbar(\omega - \omega_{nn'} + i/(\tau)_{nn'})}e^{-i\omega t} + \text{c.c.}.
\end{aligned} \tag{7.26}$$

It is found from Eq. (7.26) that the MDC oscillates and at the same time decays in an exponential way as $\exp(-t/\tau)$. After a long time the initial oscillation will die out and the driven Maxwell displacement oscillates as in Eq. (7.26) at the imposed light frequency, but not exactly in phase with the light. However, the time scale of the experiment is much longer than the period of oscillation, which leads to a zero averaged MDC of the first order over the material time spent in the experiment. Thus the second order, i.e. the nonlinear part of the transition dipole moment has to be considered.

In a similar way, the second-order part can be obtained using Eq. (7.23)

$$
\begin{aligned}
\boldsymbol{\mu}_{nn'}^{(2)}(\omega_j + \omega_k) =\ & \frac{1}{\hbar(\omega_j - \omega_k - \omega_{nn'} + i/(\tau)_{nn'})} \\
& \times \sum_{n''} \{ [H^{int}(\omega_j)]_{nn''} \boldsymbol{\mu}_{n''n'}^{(1)}(\omega_k) - \boldsymbol{\mu}_{nn''}^{(1)}(\omega_k)[H^{int}(\omega_j)]_{n''n'} \\
& + [H^{int}(\omega_k)]_{nn''} \boldsymbol{\mu}_{n''n'}^{(1)}(\omega_j) - \boldsymbol{\mu}_{nn''}^{(1)}(\omega_j)[H^{int}(\omega_k)]_{n''n'} \}. \quad (7.27)
\end{aligned}
$$

The above equation may be used to explain the *trans-cis* photoisomerization in monolayers. In the *trans* $\rightarrow$ *cis* process, $\omega_j \rightarrow \omega$ and $\omega_k \rightarrow -\omega_k$, while $\omega_k \rightarrow \omega$ and $\omega_j \rightarrow -\omega_j$ represents the *cis$\rightarrow$trans* process. Furthermore, it is obvious from Eq. (7.27) that the second order oscillates at $\omega - \omega_k$ or $\omega + \omega_k$, a frequency different from the first order. Thus it is predictable that the average of the displacement current over a full period $2\pi/\omega$ will not be zero. A simple calculation of Eq. (7.27) similar to Eq. (7.26) gives the nonvanishing MDC. This argues that the MDC generation in the experiment as carried out by the authors [27] is second order.

The explanation for the photoizomerization MDC in azobenzene monolayers can be more clearly depicted by a three-state process shown in Fig. 7.5. The energy levels include the *trans* state $| n \rangle$, with the two nitrogen-benzene bonds parallel, the intermediate state $| m \rangle$ corresponding to the transient state with a broken π-bond, and the *cis* state $| n' \rangle$ with the two nitrogen-benzene bonds and the N=N bond coplanar. The conversion between *trans* and *cis* states must involve the breaking of the π-bond. As the ultraviolet light $\omega = \omega_j$ (λ_1 in Fig. 3.17) is imposed, the molecules in the monolayers are excited from the *trans* state $| n \rangle$ to an intermediate state $| m \rangle$, and then return to a new lower state, i.e., the *cis* state $| n' \rangle$. While the visible light $\omega = \omega_k$ (λ_2 in Fig. 3.17) is imposed on the films, the molecules experience an inversion process. Both processes create induced charges given by Eq. (7.27) and generate MDC. In addition, the realistic process of photoisomerization involves many electrons rather than just one, and when the equilibrium is established for the transition between the $| n \rangle$ and $| n' \rangle$ state, the MDC in both processes decays to zero, as shown in Fig. 3.17. [27] A classical analysis using the rate equation gives the result. [29] Another influence on MDC in the *trans$\rightarrow$ cis* region as well as the *cis$\rightarrow$trans* region may be the experimental precision, as the light used in Fig. 3.16 is not strictly monochromatic.

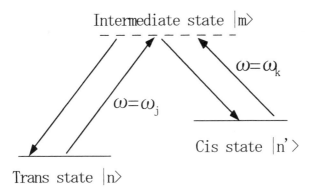

Figure 7.5 A three-state model for *trans-cis* photoisomerization in monolayer films. It is shown that such a displacement current generation is a second-order one.

7.6 Summary

In summary, the general expression of the relation between macroscopic SOS $\chi^{(2)}$ and molecular SOS $\alpha^{(2)}$ as well as their orientational order parameters for a monolayer at the liquid-air interface has been obtained. Based on the result, the chiral and nonchiral terms of $\chi^{(2)}$ have been distinguished, which can serve to study the SHG-CD effect for a chiral monolayer. Using the simple relation between their orientational order parameters and the molecular area, an SHG spectroscopy by compressing the monolayer to estimate the molecular SOS $\alpha^{(2)}$ was presented. A semiclassical perturbation theory explains the Maxwell displacement current characteristics arising from the photoisomerization phenomena in monolayer films. It is found that the Maxwell displacement current induced by photoirradiation is a second-order one generated by the interaction between the circularly polarized light and the permanent dipole moment as well as by the random perturbation of the system by the thermal reservoir.

References

[1] A. Ulman, *Characterization of Organic Thin Films*, Butterworth-Heimann, Boston (1995).

[2] J. Garnaes, D. K. Schwartz, R. Viswanathan, and J. A. N. Zasadzinski, *Nature*, **357** (1992) 54.

[3] M. Iwamoto and Y. Majima, *J. Chem. Phys.*, **94** (1991) 5135.

[4] Z. C. Ou-Yang, X. B. Xu, C. X. Wu, and M. Iwamoto, *Phys. Rev.*, **E59** (1999) 2105.

[5] W. Zhao, C. X. Wu, M. Iwamoto, and Z. C. Ou-Yang, *J. Chem. Phys.*, **110** (1999) 12131.

[6] Y. R. Shen, *J. Vac. Sci. Technol.*, **133** (1985) 1464.

[7] Y.R.Shen, *The principles of Nonlinear Optics*, (John Wiley and Sons,Inc, New York, 1984) Chap.2.4.

[8] T. F. Heinz, H. W. K. Tom, and Y. R. Shen, *Phys. Rev.*, **A28** (1983) 1883.

[9] Th. Rasing, Y. R. Shen, Mahn Won Kim, P. Vaint, Jr., and J. Bock, *Phys. Rev.*, **A31** (1985) 537.

[10] M. C. Goh, J. M. Hicks, K. Kemnitz, G. R. Pinto, K. Bhattacharyya, K. B. Einsenthal, and T. F. Heinz, *J. Phys. Chem.*, **92** (1988) 5074.

[11] S. G. Grubb, Mahn Won Kim, Th. Rasing, and Y. R. Shen, *Langmuir*, **4** (1988) 452.

[12] Th. Rasing, G. Berkovic, Y. R. Shen, S. G. Grubb, and M. W. Kim, *Chem. Phys. Lett.*, **130** (1986) 1.

[13] M. E. Rose, *Elementary Theory of Angular Momentum*, Wiley, New York (1957).

[14] T. Petralli-Mallow, T. M. Wong, J. D. Byers, H. I. Yee, and J. M. Hicks, *J. Phys. Chem.*, **97** (1993) 1383.

[15] T. D. Byers, H. I. Yee, T. Petralli-Mallow, and J. M. Hicks, *Phys. Rev.*, **B49** (1994) 14643.

[16] T. L. Mazely and W. M. Hethrigton III, *J. Chem. Phys.*, **86** (1987) 3640.

[17] A. Sugimura, M. Iwamoto, and Z. C. Ou-Yang, *Phys. Rev.*, **E50** (1994) 614.

[18] J. A. Giordmaine, *Phys. Rev.*, **A138** (1965) 1599.

[19] J. F. Nye, *Physics of Crystals*, Oxford (1957) p.113.

[20] M. Iwamoto, A. Sugimura, and Z. C. Ou-Yang, *Phys. Rev.*, **E54** (1996) 6537.

[21] D. L. Andrews and M. J. Harlow, *Phys. Rev.*, **A29** (1984) 2796.

[22] D. L. Andrews and I. D. Hands, *Chem. Phys.*, **213** (1996) 277.

[23] M. Iwamoto, C. X. Wu, and Z. C. Ou-Yang, *Chem. Phys. Lett.*, (2000).

[24] D. A. Kleinman, *Phys. Rev.*, **126**, 1977 (1962).

[25] C. C. Wang, *Phys. Rev.*, **178** (1969) 1457.

[26] J. Xue, C. S. Jung, and M. W. Kim, *Phys. Rev. Lett.*, **69** (1992) 474.

[27] M. Iwamoto, Y. Majima, H. Naruse, T. Noguchi, and H. Fuwa, *Nature (London)*, **353** (1991) 645.

[28] C. X. Wu, W. Zhao, and M. Iwamoto, *Chem. Phys. Lett.*, **309** (1999) 479.

[29] W. Zhao, C. X. Wu, and M. Iwamoto, *Chem. Phys. Lett.*, **312** (1999) 572.

CHAPTER 8

THERMALLY-STIMULATED CURRENT

Thermally stimulated current (TSC) measurement has been used for studying the dielectric depolarization phenomenon as well as the disordering degree of constituent polar molecules in multilayer film systems. [1, 2] The Maxwell displacement current flowing across multilayer films sandwiched between metal electrodes provides an indirect approach to detect the orientational motion of dipolar molecules in monolayers. MDC recorded during heating, and various peaks appearing at characteristic temperatures gives abundant information which can be attributed to relevant discharge mechanisms associated to the orientational change in polar molecules, detrapping of electron charges, etc. On the other hand, on the basis of Debye philosophy, [3] Tanguy and Hesto studied the depolarization effects associated with initial heating of orthphenanthrolin multilayers; [4] Jones et al. investigated the thermally stimulated discharge of alternate-layer Langmuir-Blodgett (LB) film structures; [5] Jonscher discussed thermally stimulated depolarization current (TSD) on multilayers on the basis of the so-called "universal law", and developed a new theoretical approach to TSD. [6] There is, unfortunately, a drawback in TSD theory which is based on classical mean-field theory with a relaxation-time treatment in explaining the critical phenomena of monolayer and multilayer films. Moreover, it is very difficult to obtain the reproducibility in the measurement, particularly in the TSC measurement of a one-layer film, principally due to the destruction of the film by the application of the top electrode. [7] A displacement current measuring system equipped with a thermal stimulation system was developed, [8] in which an air gap between the top electrode and the film surface is introduced. TSC across arachdic acid monolayers due to the disordering of multilayers using a metal/monolayer/air-gap/metal structure was then investigated. It is necessary to try the thermodynamics approach rather than TSD to explain the modern critical phenomena which happen in almost every TSC experiment. Here it is instructive to note that no biasing voltages is applied, which is the conventional approach in the TSD measurement, to the films of interest before the thermodynamic TSC measurement. In this chapter, under the frame work of classical mean-field theory, TSC will be calculated with a thermodynamics approach on the basis of the analyses developed by the authors, [9, 10, 11] and then the TSC of phospholipid monolayers due to the disordering of polar orientation of molecules will be discussed.

8.1 Thermally-stimulated Current

Monolayer and multilayer films exhibit various phases, [1] and it is expected that the behavior of the thermally stimulated current depends on phases. For simplicity, the TSC generated in monolayers sketched as the model shown in Fig. 8.1 will be discussed. In the figure, one monolayer with thickness h is deposited on electrode 2,

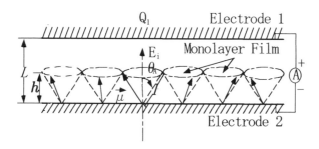

Figure 8.1 Sketch of the model for a molecule monolayer sandwiched between electrodes.

and it consists of rodlike polar molecules with a permanent dipole moment μ in the direction along the molecular long axis. The constituent molecules with a length l are orientationally distributed with a tilt angle θ away from the normal direction to the surface of metal electrode. The molecules are assumed to align in the range $0 < \theta < \theta_A$ due to effects of hard core intermolecular forces, where $\theta_A = \arcsin(\sqrt{A/A_0})$. The surface density of the constituent molecules is given by $n = 1/A$, where A is the area occupied per molecule. A_0 is the critical area occupied by a single molecule and defined as πl^2. Electrode 1 is suspended in the air and connected to electrode 2 through an ammeter. The vertical component of the effective mean dipole moment $\langle m \rangle$ of the constituent molecule without dielectric anisotropy can be expressed as [see also Eq. (4.17)] [12]

$$\langle m \rangle - m_z - \mu S + \alpha \langle E_i \rangle. \tag{8.1}$$

Here S is the first orientational order parameter of monolayers defined as $S = \langle \cos \theta \rangle$, where α is the electronic polarizability of the constituent molecules. $\langle E_i \rangle$ is the mean depolarization electric field applied to the constituent molecule in a direction perpendicular to the monolayer surface from electrode 2 to the air, which is given by

$$\langle E_i \rangle = -gn^{3/2}\langle m \rangle, \tag{8.2}$$

where g(> 0) is the interaction constant working on the constituent molecules and can be calculated depending on the arrangement of the constituent molecules and given by $g = A_0^{3/2} g_z$ (g_z is given by Eq. (2.21)). [13] The charge Q_1 is induced on electrode 1 due to the orientational distribution of the constituent polar molecules. Assuming that monolayers are heated at a constant heating rate β, a Maxwell displacement current is generated by the change of the induced charge Q_1: [11]

$$\begin{aligned} I &= -dQ_1/dt \\ &= \frac{\beta n B}{L} \frac{d\langle m \rangle}{dT}, \end{aligned} \tag{8.3}$$

where B is the working area of electrode 1 and L is the distance between electrodes 1 and 2. Using Eqs. (8.1)-(8.3), $\langle m \rangle$ and current I are obtained as follows:

$$\langle m \rangle = \frac{\mu S}{\epsilon_\infty} \tag{8.4}$$

and

$$I = \beta \frac{n B \mu}{d \epsilon_\infty} \frac{dS}{dT}, \tag{8.5}$$

with $\epsilon_\infty = 1 + \alpha g n^{3/2}$. As in Ref. [10], the dielectric constant ϵ_S of the monolayers is [11]

$$\epsilon_S = \frac{\epsilon_\infty S_0}{S}, \tag{8.6}$$

where S_0 is the order parameter in the zero field. Substituting Eq. (8.6) into Eq. (8.5), the current is obtained as:

$$I = \beta \frac{n B}{L} \mu S_0 \frac{d}{dT} \left(\frac{1}{\epsilon_S} \right). \tag{8.7}$$

It is interesting to note here that current I in Y-type multilayer films with odd number of layers is given by the same Eqs. (8.5) and (8.7), because the permanent dipole moment is canceled out due to the alternating deposition of monolayers with opposite direction layer by layer in the Langmuir-Blodgett technique, and only the contribution of one layer remains as the induced charge on electrode 1.

8.2 Depolarization due to Thermal Stimulation

Because of the axial symmetry, the molecules are considered to be restricted within the range $0 < \theta < \theta_A$ under the mean-field approximation, and the orientational order parameter S is expressed as

$$S = \int_0^{\theta_A} \cos \theta \frac{e^{-W(\theta)/kT}}{Z} \sin \theta d\theta, \tag{8.8}$$

where Z is the partition function:

$$Z = \int_0^{\theta_A} e^{-W(\theta)/kT} \sin \theta d\theta. \tag{8.9}$$

Here k is the Boltzmann constant and T the absolute temperature. $W(\theta)$ is the effective self-consistent field due to the interaction of molecules. In a zero field, S becomes Eq. (4.16). Under the mean field approximation, constituent rod-like molecules freely rotate in the range of tilt angle $0 < \theta < \theta_A$ under some mean local electric field and the interaction energy can be written as the form $W = W_0 \cos \theta$. Substituting this into Eq. (8.8), the same result is obtained as Eq. (5.22)

$$S = \frac{e^t - \cos \theta_A e^{t \cos \theta_A}}{e^t - e^{t \cos \theta_A}} - \frac{1}{t}, \tag{8.10}$$

where $t = -W_0/kT$. S is a function which increases monotonously with t. In the limit t is infinity ($t \to \infty$) S approaches 1, whereas it approaches $S = \cos \theta_A$ in the limit $t \to -\infty$. Substituting the equation above into Eq. (8.5), we obtain the general TSC of $W_0 \cos \theta$-form interaction becomes:

$$I = \beta \frac{nB\mu}{d\epsilon_\infty} [-\frac{(1 - \cos \theta_A)^2 e^{t(\cos \theta_A - 1)}}{(1 - e^{t(\cos \theta_A - 1)})^2} + \frac{1}{t^2}] \frac{dt}{dT}. \tag{8.11}$$

Eqs. (8.6), (8.10), and (8.11) will now be used to discuss the TSC mechanism.

8.2.1 Liquid Phase

In the region when $T > T_C$, the intermolecular interaction, which is the main contributor in the highly oriented phase($T < T_C$), is negligible in the fluid phase compared to the other interactions. Thus under a mean-field approximation the interaction energy can be obtained:

$$W(\theta) = -\mu \langle E_i \rangle \cos \theta. \tag{8.12}$$

It is instructive here to discuss the case for $|t| < 1$, in which S is written approximately as

$$\begin{aligned} S &= S_0 + \frac{t(1 - \cos \theta_A)^2}{12} \\ &= S_0(1 - \frac{R}{T}), \end{aligned} \tag{8.13}$$

with $R = -gn^{3/2}\mu^2(1 - \cos \theta_A)^2/12k(1 + \alpha gn^{3/2})$. On the other hand, the average dipole moment $\langle m \rangle$ is given approximately by

$$\langle m \rangle = \frac{\mu S_0}{\epsilon_S}, \tag{8.14}$$

using Eqs. (8.4) and (8.6). Here ϵ_S is the relative dielectric constant of the monolayer written as

$$\epsilon_S = \epsilon_\infty(1 + \frac{R}{T}). \tag{8.15}$$

Therefore, substituting Eqs. (8.14) and (8.15) into Eqs. (8.3) and (8.5), the TSC current I is obtained:

$$I = K(T + R)^{-2} \tag{8.16}$$

with $K = \beta R \mu S_0 n B / L \epsilon_\infty$. S, ϵ_S and I are plotted by solid lines as a function of T in Fig. 8.2. As can be seen in the figure, in the region $T > T_C$, the order parameter S gradually increases and approaches S_0 as the temperature increases, whereas the dielectric constant $\epsilon_S - \epsilon_\infty$ decreases. Current I flows in the positive direction–that is, in the direction from electrode 1 to electrode 2 through an ammeter (see Fig. 8.1)–and gradually decreases. It is instructive here to note that a linear relationship between $\epsilon_S - \epsilon_\infty$ and $1/T$ represents the characteristics of normal dielectric materials. [12] Further, it is interesting to calculate S, ϵ_S and I in the case $\langle E_i \rangle$, which is reliable when electrode 1 is in contact with monolayers on electrode 2. [14] The results are plotted by dotted lines in Fig. 8.2. As can be seen in the figure, $S = S_0$ and $I = 0$ in the region $T > T_C$. Judging from Fig. 8.2, it may be concluded that the internal electric field $\langle E_i \rangle$ works to suppress the polar orientational ordering, that is, the field works to decrease S, whereas the increase in temperature tends to overcome the suppression.

8.2.2 Liquid-crystalline Phase

In the liquid-crystalline phase when $T < T_C$, interdipole attractive forces such as the Keesom interaction, which is inversely proportional to temperature T, and other temperature-independent interactions are working among the constituent molecules. [15] As a result, in such a case of the so-called liquid phase, when the temperature is relatively high, interaction among molecules is small compared with interaction between molecules and the interface, and in the liquid-crystalline phase, molecules are highly oriented, and interaction among molecules is the dominant factor and should be considered. In the region $T < T_C$, for one molecule, because of the system's space-inversion symmetry, it is postulated that the interaction as a mean-field effect, and the Keesom-form interaction is temperature dependent:

$$W(\theta, T) = -V_0\left(\frac{T_C}{T} - \Gamma\right)\cos\theta - \mu\langle E_i \rangle \cos\theta. \tag{8.17}$$

Here V_0 is the magnitude of the interaction, Γ the transition order and $\langle E_i \rangle$ the mean field which is expressible as fS ($f = -gn^{3/2}\mu/\epsilon_\infty < 0$) according to Eqs. (8.2) and (8.4). Thus t is

$$\begin{aligned} t &= \frac{V(T) + \mu\langle E_i \rangle}{kT} \\ &= \frac{V_0}{kT}\left(\frac{T_C}{T} - \Gamma + \frac{\mu f}{V_0}S\right). \end{aligned} \tag{8.18}$$

In case $|V_0| >> kT >> |\mu f|$, i.e., the mean field effect is negligible, considering Eqs. (8.6), (8.10), (8.11), and (8.18), S, ϵ_S and TSC of region $T < T_C$ may be

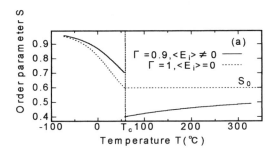

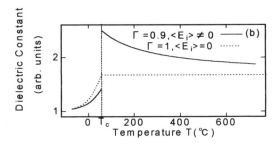

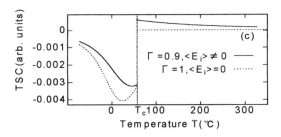

Figure 8.2 Theoretical calculation. Dotted lines, with $\Gamma = 1$, $\cos\theta_A = 0.2$, $T_C = 330$ K and $V_0 = 20kT_C$ for $\langle E_i \rangle = 0$. Thin lines with $\Gamma = 0.9$, $\cos\theta_A = 0.2$, $T_C = 330$ K and $V_0 = 20kT_C$ for $\langle E_i \rangle \neq 0$. (a) Thermally stimulated order parameter S. (b) Temperature-dependent dielectric constant ϵ. (c) TSC.

expressed as:

$$S = \frac{1 - \cos\theta_A e^{(\cos\theta_A - 1)\frac{V_0}{kT}(\frac{T_C}{T} - \Gamma)}}{1 - e^{(\cos\theta_A - 1)\frac{V_0}{kT}(\frac{T_C}{T} - \Gamma)}} - \frac{1}{\frac{V_0}{kT}(\frac{T_C}{T} - \Gamma)} \tag{8.19}$$

$$\epsilon_S = \frac{\epsilon_\infty S_0}{S} \tag{8.20}$$

$$I = \beta\frac{nB\mu}{d\epsilon_\infty}[-\frac{(1 - \cos\theta_A)^2 e^{\frac{V_0}{kT}(\frac{T_C}{T} - \Gamma)(\cos\theta_A - 1)}}{(1 - e^{\frac{V_0}{kT}(\frac{T_C}{T} - \Gamma)(\cos\theta_A - 1)})^2}$$
$$+ \frac{1}{\frac{V_0}{kT}(\frac{T_C}{T} - \Gamma)}](-\frac{2V_0T_C}{kT^2} + \frac{\Gamma V_0}{kT}). \tag{8.21}$$

It is easy to estimate that at room temperature, the relation $|\mu f| << kT$ is satisfied for the real monolayer film systems, e.g., $L - \alpha$-dimyristoyl, phosphatidylcholine (DMPC) films, as will be disscussed in section 8.4. Eqs. (8.19), (8.20), and (8.21) are depicted in Fig. 8.2. As can be seen in the figure, S decreases as temperature increases possibly because the inter-molecular attractive interaction effect is weaken as the temperature increases, whereas ϵ_S increases and current I flows in the negative direction as the temperature increases. It is obvious from Fig. 8.2, that the transition temperature

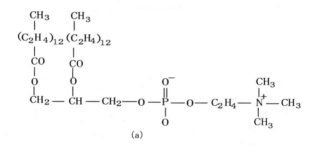

(a)

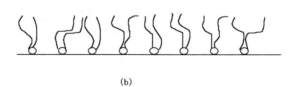

(b)

Figure 8.3 (a) Molecular structure of $L - \alpha$-dimyristoyl, phosphatidylcholine (DMPC). (b) Monolayer structure consisting of DMPC molecules.

is easily determined from the Maxwell displacement current (MDC) measurement because the direction of current changes at the transition temperature T_C.

While in case $|V_0| << |\mu f| << kT$, $t = t' + \frac{\mu f}{kT}S$. Under a simple approximation, Eq. (8.10) leads to

$$
\begin{aligned}
S &= \frac{S(t')}{1 - (\frac{\partial S}{\partial t'})\frac{\mu f}{kT}} \\
&= S(t') + (\frac{\partial S}{\partial t'})S(t')\frac{\mu f}{kT},
\end{aligned}
\tag{8.22}
$$

where $S(t')$ is expressed as Eq. (8.10) and

$$
t' = \frac{V_0}{kT}(\frac{T_C}{T} - \Gamma).
$$

The dielectric constant and Maxwell displacement current can also be obtained by Eq. (8.6) and Eq. (8.7).

8.3 TSC Experiment

The experimental system was the same as reported by the authors. [8] The molecular structure of $L - \alpha$-dimyristoyl, phosphatidylcholine (DMPC) is shown in Fig. 8.3. It is composed of a main chain with two tails. The TSC of a three-layer $L - \alpha$-dimyristoyl, phosphatidylcholine (DMPC) film was measured at a heating rate β of 0.2 °C/sec and plotted in Fig. 8.4(a). Here the film was prepared with the conventional vertical dipping method in the LB technique at a surface pressure of 30 mN/m by raising and dipping the substrate through an air-water interface. The molecular area ($A = 1/n$) of the deposited DMPC film was about 63 Å². It is found that the direction of current changes abruptly at a temperature of 60 °C, corresponding to the transition temperature, [15] and that this directional change in current reproducibly appears during heating by cyclic heating and cooling. In order to examine the dielectric behavior of DMPC, the capacitance of three-layer DMPC multilayers sandwiched between aluminum metal electrodes was measured by increasing the temperature as a function of frequency. The results were plotted in Fig.8.4(b). The capacitance climbes up to a maximum at 60 °C and then decreases after the transition. [16]

8.4 Phase Transition

The theoretical thermally stimulated current expressed by Eqs. (8.16) and Eq. (8.21) and plotted in Fig. 8.2(c) appears to have a shape similar to the experimental result shown in Fig. 8.4(a), where the direction of current changes at a transition temperature of 60 °C. This result suggests that the mean-field effect is negligibly small, and the relation $|\mu f| << kT << V_0$ is satisfied in the temperature range between 0 °C

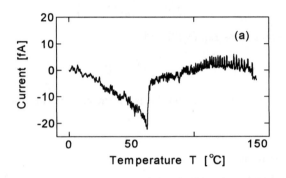

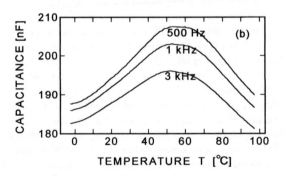

Figure 8.4 (a) Experimental TSC of DMPC ($\beta = 0.02°$ C/sec). (b) Temperature-dependent capacitance ($\beta = 0.02°$ C/sec).

and 60 °C ($=T_C$). In order to clarify this suggestion, here the value of $\mu f/kT$ is calculated. From the results of MDC-measurement of DMPC, [17] it may be assumed that the dipole moment μ is assumed to be about $0.65 \sim 0.80$ D, and the length of the DMPC molecule a is 0.944 nm. Further, as described in Sec. IV, the molecular density n $(1/A)$ is estimated to be 1.6×10^{18} m^{-2}. Thus the value of $\mu f/kT$ is estimated to be about 2.4×10^{-4} for $g = 1$, $\epsilon_\infty = 1.2$, and $T = 300$ K. Therefore it may be concluded that the relation $|\mu f| \ll kT \ll V_0$ is satisfied in the experimental results. As can be seen from the calculation results plotted in Figs. 8.2(a) and 8.2(c), TSC becomes smaller as $\langle E_i \rangle$ decreases. In the experiment, μf is very small in comparison with kT in the temperature range $T > T_C$. As a result, the experimental TSC is very small in the range $T > T_C$ [see Fig. 8.4(a)], as expected. It is still not clear why the theoretical calculation leads to a small cusp at a temperature lower than T_C, which becomes obscure and shifts to a lower temperature as V_0 decreases. A consideration of relaxation effect is required for a profound understanding of the transition in the experimental TSC curves.

As for the capacitance, the total capacitance C is expressible as

$$\frac{1}{C} = \frac{1}{C_0} + \frac{1}{C_S}, \tag{8.23}$$

where C_0 is the capacitance of the native aluminum oxide layer formed on the electrodes, and C_S is the capacitance of the film. Thus from Eq. (8.20) and the expression above, the same capacitance appearance as in Fig. 8.4 can easily be obtained, because the capacitance of the native aluminum oxidelayers is independent of the temperature in the range employed in the TSC measurement. [14]

8.5 Thermodynamics Approach to Monolayers

An alternative model has been described, which is expected to be more appropriate for explaining TSC phenomena. The experimental TSC shows a sudden direction change at the transition point, which could be employed to determine the critical transition temperature T_C. The conventional relaxation-time treatment, which is based on nonequilibruim thermostatistics, has been studied for a couple of decades; however, there are still some difficulties in revealing the phase-transition phenomena of monolayers. The present model reveals that, as temperature increases, the monolayer experiences an ordering liquid-crystalline state and a disordering liquid state. The intermolecular interactions, which are temperature-dependent, dominate the TSC generation in the liquid-crystalline state. Γ is an index representing the order of phase transition as that in the last chapter, i.e.,

$$\begin{cases} 0 < \Gamma < 1 & \text{first} - \text{order transition} \\ \Gamma = 1 & \text{second} - \text{order transition}. \end{cases}$$

Criteria for a higher-order phase transition requires a further comparison between the intermolecular and molecule-interface interactions, and a consideration of the contribution of the mean internal electric field. Because a determination of transition order needs further investigation, $\Gamma \neq 1$ and $\Gamma = 1$ are set in the simulation with and without consideration of internal electric field respectively to explain the experimental appearance of the capacitance and TSC. The calculcated TSC reveals that the first-order simulation is closer to the experimental TSC than the second-order one.

8.6 Summary

A model concerning phase transition due to molecular orientational disordering in monolayers was established with a thermodynamical approach. The thermally stimulated current, dielectric constant and orientational order parameter were calculated, and experimental results on DMPC were discussed based on the model presented here. Monolayers under thermal stimulation experience a phase transition from the liquid-crystalline phase to a liquid phase. The estimation that the mean-field effect is negligible in the liquid-crystalline phase shows that two different kinds of interactions dominate the two phases respectively. An abrupt change of MDC direction occurred at the critical temperature due to the sudden change of orientational ordering, though a cusp at a temperature lower than T_C in Fig. 8.2(c) appears depending on the magnitude of the Keesom interaction V_0. The phase transition of DMPC is close to a first-order one. The change of dielectric constant can also be discussed by such a model, as long as the orientational order parameter is obtained. By further consideration of the relaxation effect, though complicated, a smoother TSC curve can be expected. Finally it should be noted that the model discussed above relies upon the assumption that molecules are quick enough to respond as temperature increases. Its chief advantage lies in that it can explain the TSC phenomena without calculating the complicated relaxation time. [18]

References

[1] G. L. Graines, Jr., *Insoluble Monolayers at Liquid-Gas Interface*, Interscience, New York (1965).

[2] R. Chen and Y. Kirsh, *Analysis to Thermally Stimulated Process*, Pergamon, Oxford (1981); T. Hino, *IEEE*, **EI** − **21** (1986) 1007.

[3] P. Debye, *Polar Molecules*, Dover, New York (1929).

[4] J. Tanguy and P. Hesto, *Thin Solid Films*, **21** (1974) 129.

[5] C. A. Jones, M. C. Petty, G. Davies, and J. Yarwood, *J. Phys. D*, **21** (1988) 95.

[6] A. K. Jonscher, *J. Phys. D*, **24** (1991) 1633.

[7] M. Iwamoto, T. Kubota, and M. Sekine, *J. Phys. D*, **23** (1990) 575 and the references cited therein.

[8] T. Kubota and M. Iwamoto, *Rev. Sci. Instrum.*, **64** (1993) 2627.

[9] M. Iwamoto, C. X. Wu, and W. Y. Kim, *Phys. Rev.*, **B54** (1996) 8191.

[10] A. Sugimura, M. Iwamoto, and Z. C. Ou-Yang, *Phys. Rev.*, **E50** (1994) 614.

[11] M. Iwamoto, Y. Mizutani, and A. Sugimura, *Phys. Rev.*, **B54** (1996) 8186.

[12] H. Fröhlich, *Theory of Dielectrics*, Clarendon, Oxford (1958).

[13] D. M. Taylor and G. F. Bayes, *Phys. Rev.*, **E49** (1994) 1439.

[14] C. X. Wu, Y. Mizutani, and M. Iwamoto, *Jpn. J. Appl. Phys.*, **36** (1997) 222.

[15] J. N. Israelachvili, *Intermolecular and Surface Forces*, Academic, London (1985).

[16] D. Marsh, *CRC Handbook of Lipid Bilayers*, CRC, Boston (1990).

[17] K. S. Lee and M. Iwamoto, *J. Colloide Interf. Sci.*, **177** (1996) 414.

[18] M. Iwamoto and C. X. Wu, *Phys. Rev.*, **E54** (1996) 6603.

CHAPTER 9

ELECTRONIC PROPERTIES AT MIM INTERFACES

9.1 Tunneling Current and Electronic Device Applications

One of the most important and specific applications of organic ultra-thin films in the field of electronics is to use organic thin films as an electrically insulating tunneling barrier. [1, 2] Thus there has been a growing interest in the preparation method of high-quality organic ultra-thin films. Unfortunately, it is still not an easy task to fabricate tunneling junctions using organic thin films with a film thickness less than several ten nano-meters, owing to the presence of pinholes in the films and the difficulty in the deposition of top electrodes onto the prepared films without destroying their textures. However, during the last ten years, preparation techniques have greatly advanced and various new organic materials have been synthesized. Among them are polyimides and phtalocyanine derivatives. For example, polyimide Langmuir-Blodgett (PI LB) films with a monolayer thickness of 0.4 nm have been successfully prepared by a precursor method coupled with the conventional LB technique. [3] The prepared PI LB films are amorphous-like and almost pinhole-free. The surface of these films is very flat. PI LB films are thermally and chemically stable up to a temperature of 400 °C. Their electrical resistance is very high, usually greater than 10^{15} Ωcm. Their electrical breakdown strength is higher than 10^7 V/cm. The PI films function as good electrical insulating barriers in metal-insulator-metal (MIM) structures, and also work as a good electrical tunneling barrier in tunnel junctions such as noble metal-barrier-super conductor and Josephson junctions. [4, 5] Briefly, the current-voltage (I-V) characteristic of Au/PI/Pb-Bi (30%) junctions show a nonlinear dependence as shown in Fig. 9.1(a), [4, 5] and the energy gap Δ/e estimated from this I-V characteristic changes in a manner as predicted by Bardeen-Cooper-Schrieffer (BCS) theory [see Fig. 9.1(b)]. Josephson junctions with a structure of Nb/Au/PI/Pb-Bi show a typical I-V characteristic observed in so-called superconductor-insulator-superconductor (SIS) Josephson junctions, and they are very effective in the detection of microwaves; [4, 5] that is, the normal step structure ruled by the Josephson frequency condition $V = nh/2ef(n = 1, 2, 3...)$ is clearly seen in the I-V characteristic at a temperature of 4.2 K in the presence of microwaves (see Fig. 9.2). Here, f is the applied microwave frequency, and h is the Planck constant. It was suggested that the Josephson junction is able to detect microwaves with a high frequency, e.g., 0.5 THz.

Another important research field using ultrathin films as a tunneling barrier is in building up molecular rectifying junctions, electron resonance tunneling devices, and single electron tunneling devices. [6, 8, 9] For this purpose, well-defined structures incorporating pinhole free ultra-thin films possessing functionalized molecules between two metals must be prepared. Using a two-layer system consisting of octasubstituted pal-

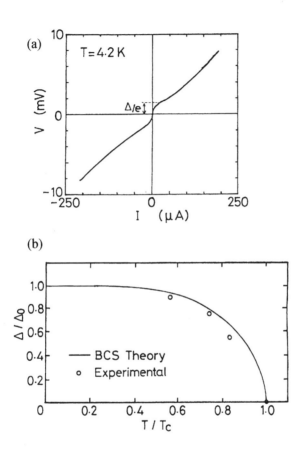

Figure 9.1 (a) A typical I-V characteristic of Au/PI/(Pb-Bi) junction at a temperature of 4.2 K, where a 27-layer PI LB film was used as a tunneling spacer. (b) Relationships between the energy gap Δ/Δ_0 and the temperature T/T_c. Δ_0 is the theoretical energy gap of a Pb-Bi alloy at a temperature of zero, and T_c is the critical temperature of the alloy. Solid line is the theoretical curve calculated from the BCS theory.

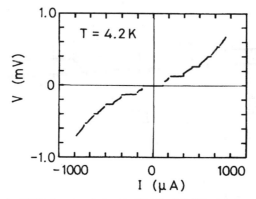

Figure 9.2 A typical I-V characteristic of a Nb/Au/PI/(Pb-Bi) junction in the presence of microwave with a frequency of 70 GHz.

ladiumphthalocyanine (PcPd) and perylene-tetra-carboxyldiimide derivative such as PTCDI-Spent and PTCDI-OEt LB films, new electronic molecular diodes have been created on the basis of the asymmetric tunneling through molecular states, [10, 11] such as Highest Occupied Molecular Orbital (HOMO) and Lowest Unoccupied Molecular Orbital (LUMO) states of these molecules. As such, it is a very important method to make clear the electron transport mechanism via functionalized molecules in artificially arranged multilayer films. Using PI LB films containing porphyrin (PORPI), elastic electron tunneling process and inelastic tunneling process of ultarathin films via porphyrin molecules have been examined, [12] and the increase of electron tunneling current due to the excitation of electronic transition in molecular states in Q bands of porphyrins has been revealed in the inelastic tunneling spectra of junctions with a structure of Au/PI/PORPI/PI/Pb (or Pb-Bi, Au) at a temperature of 4.2 K. Similar experiments were carried out for Au/PI/Rhodamine-dendorimer/PI/Au (or Al) junctions using rhodamine-dendorimer, [13] and the creation of a step structure in the I-V characteristic has been revealed as shown in Fig. 9.3. This characteristic is very similar to that seen in the so-called Coulomb staircase. [6, 7] The possibility of single electron tunneling process via rhodamine molecule as a quantum dot has been strongly suggested. From the viewpoint of dielectric physics and electrical insulation engineering, it is essential to clarify the electron injection and transport mechanism of organic monolayer. At least, the elastic and inelastic tunneling processes of organic thin films, taking into account the surface electron density of states, energy states of the constituent molecules, and other important properties need to be clarified.

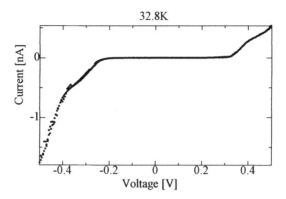

Figure 9.3 A typical I-V characteristic of Au/PI/Rhodamine-dendorimer/PI/Al junctions using rhodamine-dendorimer.

9.2 Nanometric Interfacial Electrostatic Phenomena in Ultrathin Films

Many investigations have been carried out to build up tunnel junctions and molecular rectifying junctions using organic materials with the hope of observing novel and useful electrical and optical properties. [14] For this purpose, it is essential to construct well-defined structures using pinhole free ultra-thin organic films. [1, 14, 15] Many organic materials including electrical insulating materials, [16] semi-conductor materials, [1, 2] electrical conductors [17] have been developed, and the current-voltage (I-V) characteristics of the junctions using these materials have been examined. However these are not sufficient. The interfacial electronic phenomena occurring at the metal/film and film/film interface must be clarified for a complete understanding of the operation of these junctions, possibly because the I-V characteristics of these junctions will be ruled by the nano-metric interfacial phenomena. [18, 19] Similarly, the understanding of the interfacial phenomena is a key for further development of commercialized organic material application such as electroluminescent device, [20] liquid crystal display, [21] photocopying, [22] and others. Beyond that, the study on the electrostatic interfacial phenomena is obviously important to the fields of future electrostatics and electrical insulation engineering, [23] and this study has been a continuous subject since the discovery of contact electrification phenomena. [24, 25]

For the understanding of the interfacial electrostatic phenomena, it is very helpful to use ultra-thin films whose thickness is less than the electrostatic double layer, and then to gain information on the distribution of the electronic density of states as well as the space charge distribution of excess charges in the films. [26] LB films will be suitable, because they can be prepared onto solid substrates by the layer-by-layer deposition with an order of monolayer thickness. [27] For example, electrically insulating PI LB films [3] are favorable from the viewpoint of electrical engineering. [28, 29] On the one hand, semi-conductor phthalocyanine films are interesting in electronics, [30, 31, 32] because they can be used in electronic devices such as solar cells, gas sensors, photo-conductors etc. Similarly, many electrically conductive LB films to be used as electronic wires in monolayer film devices are interesting. [17]

Obviously, the method to gain information on the nanometric interfacial electrostatic phenomena needs to be established. In the field of electrical insulation engineering, several techniques have been developed to investigate the charge distribution in insulating films. [33, 34] Among them are the heat-pulse-propagation technique, [35] the pressure pulse technique, [36] the electron beam method, [37] and the electric stress-pulse technique. [38] These techniques have the advantage in that the space charge distribution itself can be measured without being destroyed. Unfortunately, these measurements cannot be applied to the determination of space charge distribution in ultra-thin films because the resolution of these measurements is limited by the acoustic sound velocity of the materials, and it is thus limited to the order of micrometers. However, the surface potential measurement can be employed by coupling the film preparation method although this measurement is a classical one.

9.2.1 Electrically Insulating Ultrathin Films

Before going to the discussion on the interfacial electrostatic phenomena, the origins contributing to these in LB films will be briefly discussed. These can be classified as

(1) excess electronic charges displaced from metal electrodes $(\rho(x))$,

(2) permanent dipole moment (μ) of constituent molecules,

(3) surface charges adsorbed on the film surface,

(4) extrinsic ionic charges in film, and

(5) others.

Among these, the surface charges could be removed prior to measurement. For example, water molecules adsorbed on the film surface can be removed by heat treatment. Similarly, the contribution of extrinsic ionic charges embedded in films by using ultrapure water in the film deposition can be removed. Therefore, the first two origins remain to be the main contributors to the establishment of surface potential if films are placed in the dark without being subjected to such as photoillumination. Since excess electronic charges with a space charge density $\rho(x)$ are displaced into LB films at a distance x from metal electrodes, electric flux diverging from the excess charges in films falls on the metal electrodes (see Fig. 9.4). Taking these into account, the relations expressing the potential V_s (surface potential) across film in the following

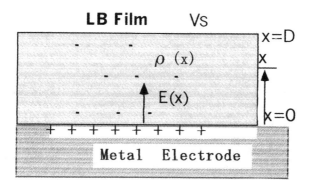

Figure 9.4 Model of excessive electronic charges displaced from metal electrodes.

sections are given by the sum of the potential by displaced electronic charges and permanent dipole moments.

Multilayer films are most commonly used in the experiments of electrostatic phenomena. They can be prepared by transferring monolayers on water surface onto substrate layer-by-layer using the LB technique. The spontaneous polarization P_0 induced across LB films depends on the preparation process. It should be noted here that at the metal/organic interface the constituent molecules of the monolayer closest to the metal surface point toward the organic monolayer side due to the presence of the symmetry-breaking interface. The spontaneous polarization exists only in films consisting of molecules with a permanent dipole moment μ, depending on the film structure. It can be expressed as

$$P_0 = nNm_z \quad \text{for Z} - \text{type film,}$$
$$P_0 = -nNm_z \quad \text{for X} - \text{type film,}$$
$$P_0 = Nm_z \quad \text{for Y} - \text{type film with odd number of deposited layers,}$$

and

$$P_0 = 0 \quad \text{for Y} - \text{type film with even number of deposited layers.}$$

Here n is the number of deposited layers, N is the surface density of molecules and it is given by $1/A$ (A: molecular area), and m_z is the vertical component of dipole moment of the constituent molecule, i.e., dipole moment perpendicular to the material surface and is given by Eq. (4.17)

In order to clarify the origin of space charges, the characteristic behavior of the potentials arising from the origins (1) and (2) listed earlier must be known. If the origin is the permanent dipole moment of the constituent molecules, the potential built across films should not depend on the variety of metal electrodes, and the surface potential should increase in proportion to the number of deposited layers for Z-type films, showing a zigzag dependence for Y-(or X-)type film. On the other hand, if the origin is excess charges displaced from metals, the surface potential will clearly depend on the variety of metal electrodes, and will be independent of the type of film, and the potential saturates as the number of layers increases possibly because the distance the electronic charges displaced from electrodes can reach may be limited to the vicinity of the interface. As will be shown later in this chapter, the main charge contribution of charges in PI and phthalocyanine LB films is the excess charges displaced from metals.

PI, with its molecular structure shown in Fig. 3.22(b), has a large electron affinity

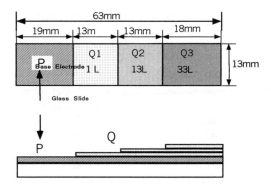

Figure 9.5 Electrode configuration of the samples used in the surface potential measurement.

and thus has a strong tendency to accept electrons. Figure 9.5 shows the electrode configuration of the sample used for the surface potential measurement. That is, polyimide LB films were deposited onto Au, Cr and Al base-electrodes. The resulting PI LB films were heat-treated for more than one hour at a temperature of 150 °C in a vacuum of the order of 10^{-6} torr, in order to remove water molecules adsorbed on the surface of PI samples and excess charges generated inside the samples. After that, the surface potential of PI LB films (at position Q indicated in Fig. 9.5) was measured with reference to the potential of the clean base metal electrodes (at position P in Fig. 9.5). Temperature dependence of the surface potential was measured at an interval of 25 °C between -100 °C and 150 °C after keeping each temperature for more than one hour. Figures 9.6(a), (b) and (c) show the relationship between the potential across PI LB films on Au, Cr and Al electrodes and the number of deposited layers at various temperatures. [39, 40] The surface potentials gradually decrease as the number of deposited layers increases, and then reach a constant saturated potential at the number of 20-50 layers. These results indicate that PI LB films acquire electrons from

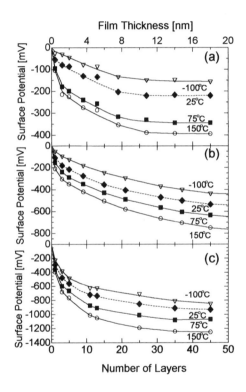

Figure 9.6 Relationship between the surface potential of PI LB films and the number of deposited layers at various temperatures on (a) Au, (b) Cr and (c) Al electrode.

metal electrode and the electrostatic layer is thus formed at the metal/PI LB film interface within a range of the order of nano-meters. This is due to the fact that the potential does not show a remarkable dependence of the base metal electrodes if the main contribution comes from the alignment of the constituent permanent dipoles in LB films. [41, 42] Further it is instructive to note here that the surface potential of PI

LB films deposited on epitaxially grown Au electrodes shows a similar dependence with that of PI LB films deposited on Au evaporated electrodes. [43] Thus the effect of the diffusion of base metal Au atoms is negligible. The value of surface potential in PI LB films shifts negatively as the temperature increases, indicating that the tendency of PI to accept electrons becomes stronger as the temperature increases.

Since excess charges are displaced into PI LB films from metal electrodes, electric flux diverging from the excess charges in PI LB films falls on the metal electrodes (see Fig. 9.4). Therefore the surface potential Vs across PI LB films is given by [44]

$$V_s = \int_0^D \frac{x\rho(x)}{\epsilon_0\epsilon_r} dx. \tag{9.1}$$

Here, ϵ_r (= 3) is the relative dielectric constant of PI, D is the film thickness, x is the distance from metal electrode, and $\rho(x)$ is the space charge density at position x. Differentiating surface potential V_s with respect to the film thickness D gives a quantity proportional to the space distribution of charges $\rho(D')$; that is, $\rho(D')$ is obtained experimentally by measuring the change in the surface potential ΔV_s with

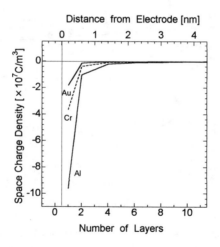

Figure 9.7 Space charge distribution in polyimide (PI) LB films.

the increment of thickness of one-layer ΔD as follows;

$$\rho(D') = \frac{\epsilon_0\epsilon_r}{D'} \cdot \frac{\Delta V_s}{\Delta D}. \tag{9.2}$$

In the calculation, the displaced electrons are assumed to locate at position $D' = (n - 1/2)\Delta D (n = 1, 2, 3...N)$, where ΔD is the monolayer thickness and n is the number of deposited layers.

Figure 9.7 shows an example of the space charge density $\rho(D)$ in PI LB films at a temperature of 25°C, calculated using Fig. 9.6 in combination with Eq. (9.2). The space charge density decreases steeply as the number of layers increases. Most of excess charges exist in PI LB films within the distance of 4 nm from electrodes. About 1 to 10% of monomer units of PI accept electrons from metal electrodes in this region, where the density of PI molecule unit is about 3×10^{27} m^{-3}. [3]

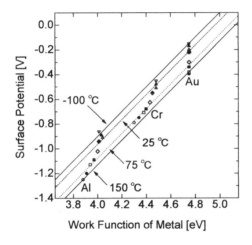

Figure 9.8 Relationship between the work function of metals and the saturated surface potential of PI LB films.

Figure 9.8 shows the relationship between the work function of metals and the saturated surface potential of PI. A linear relationship with a slope of unity is observed between them, indicating that the difference in the saturated potential of PI LB films deposited on various electrodes coincides with the difference in the work function of these electrodes. Here the work function of Au evaporated electrode was estimated to be 4.75 eV from the ultraviolet photo-emission spectroscopy. Work functions of Cr and Al electrodes were estimated using Au electrode as the reference by means of the contact potential method. [45] A linear relationship observed in Fig. 9.8 indicates that a thermodynamics equilibrium is established at the metal/PI LB film interface at each temperature. In other words, electronic charges are transferred at the interface between PI films and metals until the surface Fermi level of PI and the Fermi level of metal electrodes coincide, as shown in Fig. 9.9(a). [46, 47, 48] Therefore, the electronic states of PI whose electronic energy is higher than the Fermi level of metal can donate electrons to metal if the states are filled with electrons before electrification (electron donor states n_D). By contrast, the electronic states of PI whose electronic energy is lower than the Fermi level of metal can accept electrons from metal if the states are empty (electron acceptor states n_A). As shown in Fig. 9.5, PI LB films are charged negatively, indicating electron-acceptor states take a dominant part in the charge exchange phenomena at the metal/PI LB film interface. Therefore, the space charge density $\rho(x)$ is approximately written as: [39, 46]

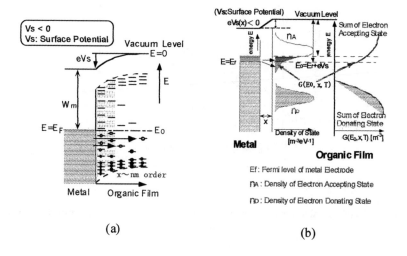

(a)   (b)

Figure 9.9 (a) Schematic illustration of the spatial charge distribution in LB films. (b) Distribution of the electronic density of states at the metal/LB film interface.

$$\rho(x) \approx \int_{-\infty}^{+\infty} -e \cdot n_A(E, x, T) f(E - eV_s) dE. \qquad (9.3)$$

Here, $f(E)$ is a Fermi-Dirac distribution function defined as

$$f(E) = \frac{1}{1 + \exp[(E + W_m)/kT]}, \tag{9.4}$$

and e is electron charge. In Eqs. (9.3) and (9.4), E represents the depth of the energy measured from the vacuum level (V.L.) [see Fig. 9.9(a)], where $E = 0$. It should be noted here that the electronic states n_A depend on energy E, position x, and temperature T [see Fig. 9.9(b)]. W_m is the work function of metal electrode. Fermi level of the metal electrode locates at $-W_m + eV_s$ from vacuum level at $x = 0$. V_s is the electrostatic potential at position x in PI LB film. The electronic surface states

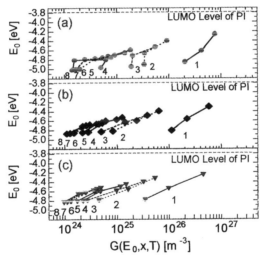

Figure 9.10 The distribution of electronic density of states and the sum of density of state $G(E_0, x, T)$. Relationship between $G(E_0, x, T)$ for PI LB films and the energy E_0 of PI at (a) 150°C (b) 25°C, and (c) -100°C.

are distributed with over several hundred milli-volts in electron energy, much larger than thermal energy kT. Therefore, $f(E)$ is approximately written as a step function varying around the Fermi level. It should be noted here that the energy E_0 just corresponds to the Fermi level of the base metal electrode measured from the V. L. in the film at position x after contact with the PI and base electrode. Because of the energy level shift due to the electrostatic potential V_s, E_0 at position x is expressed by $-W_m + eV_s$ (see Fig. 9.9(a)). Thus Eq. (9.3) becomes approximately

$$\rho(x) \approx \int_{-\infty}^{+E_0} -e \cdot n_A(E, x, T)dE. \tag{9.5}$$

The term given by $\mid \rho(x)/e \mid (G(E_0, x, T))$ represents the sum of the electron acceptor states n_A in the range of energy between $E = -\infty$ and $E = E_0$, which are occupied with electrons after electrification. It is instructive to note here that the profile of n_A does not depend on the work function of base electrode, whereas the magnitude of E_0 depends on the work function of base electrode and the electrostatic potential V_s built in films due to the electrification. Therefore, E_0 can be altered by choosing the base metal electrode material, and $G(E_0, x, T)$ is obtained experimentally for the corresponding base material as illustrated in Fig. 9.9(b).

Figure 9.10(b) shows the relationship between $G(E_0, x, T)$ and the depth of energy E_0 at room temperature (25°C), obtained from Fig. 9.5 and Eq. (9.2). Curves 1-8 represent the relationship between $G(E_0, x, T)$ and E_0 at positions $x = 0.2, 0.6, 1.0, 1.4, 1.8, 2.2, 2.6$ and 3.0 nm, respectively. As plotted in Fig. 9.10, it was found that $G(E_0, x, T)$ decreases steeply as the distance from metal increases in the range of energy -4.4 through -4.9 eV. This result indicates that the electron acceptor states locating at deep energy levels are confined within the region of 1-2 nm from the metal/PI LB film interface, because $G(E_0, x, T)$ decreases steeply at position $x = 0.6$ nm (curves 2 and 3). The order of the density of state, 10^{26} m^{-3} of $G(E_0, x, T)$ indicates that about 10

Figure 9.10(a) and (c) show the relationships between $G(E_0, x, T)$ and the depth of energy E_0 at 150°C and -100°C. It was found from Fig. 9.10 that $G(E_0, x, T)$ increases at lower energy levels as temperature increases. The distribution of the electron acceptor states is broaden as the temperature increases.

9.2.2 Semiconductor Films

Cu-tetra-tert-butyl-phthalocyanine (CuttbPc), whose chemical structure is shown in Fig. 9.11, were also examined. [49, 50, 51] Each CuttbPc molecule has one copper atom at the center of phthalocyanine rings. First, monolayers of CuttbPc spreading on a water surface were transferred onto glass substrates covered with Au, Cr, or Al evaporated electrodes by the horizontal lifting method at a surface pressure of 20 mN/m and a molecular area of 50 Å^2 at 20°C, except that the first layer which was deposited by the vertical dipping LB technique. The monolayer thickness of CuttbPc LB films was determined as 1.7 nm from the X-ray diffraction pattern. Before the surface potential measurement, all samples were heat-treated for more than one hour at a temperature of 70°C in a vacuum of the order of 10^{-6} Torr (1.33×10^{-4} Pa).

Figure 9.12(a) shows the relationship between the surface potential of CuttbPc LB films on Au electrode and the number of deposited layers (n) at various temperatures. The surface potentials gradually increase as the number of deposited layers increase for $n < 3$. In contrast, the potentials gradually decrease as the number of deposited

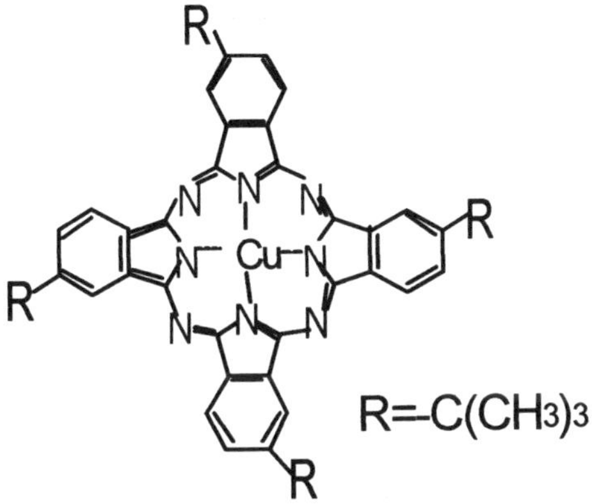

Figure 9.11 Molecular structure of Cu-tetra-tert-butyl-phthalocyanine (CuttbPc).

layers increase for $n > 3$, and the potentials reach saturated values in the range $n = 10 - 20$, indicating that the excess charges are displaced in the range within about 17-34 nm from the film/electrode interface. The saturated surface potential gradually decreases as the temperature decreases, and it changes the polarity at a temperature below 20°C. Figure 9.12(b) shows the surface potential on Cr electrode. The surface potentials gradually decrease as the number of deposited layers increase, and they finally reach saturated potentials at $n = 5$. Likewise, the surface potential gradually decreases as the temperature decreases in a manner similar to the surface potential change of CuttbPc on Au electrodes [see Fig. 9.12(a)]. Figure 9.12(c) shows the surface potential on Al electrodes. The surface potential gradually decreases as the number of deposited layers increases, and it reaches a saturated potential at around $n = 8$. Of interest is that the surface potential on Al electrodes becomes increasingly negative between 20 and 100°C. This is a major characteristic seen for CuttbPc LB films on Al electrodes, and this characteristic was not seen for CuttbPc LB films on Cr nor Au electrodes [see Figs. 9.12(a) and 9.12(b)].

As aforementioned, the surface potential across CuttbPc LB films depends on base metal electrodes as well as the number of deposited layers. Similar results were obtained for PI LB films deposited on metal electrodes as mentioned earlier, [26, 49] in which the surface potential built across PI LB films was due to the displacement of electrons from metal electrodes. Thus we may expect that electronic charge exchange phenomena occur at the metal/CuttbPc film interface in a similar way. Figure 9.13 shows the relationship between work function of metals and the saturated surface potential of CuttbPc LB films at 100, 20 and -100°C. Here, the work functions of Au, Cr and Al were 4.75 eV, 4.35 and 3.75 eV at 20°C. The work functions of Cr and

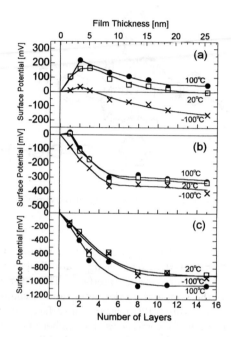

Figure 9.12 Relationship between the surface potential of CuttbPc LB film on (a) Au electrodes, (b) Cr electrodes and (c) Al electrodesand, and the number of deposited layers at various temperatures.

Al obtained by the contact method with reference to the bare-Au electrode shows the temperature dependence, and the work functions of Cr and Al were determined as 4.35 and 3.70 eV at 100°C, and 4.40 and 3.75 eV at -100°C, respectively. Linear relationships were observed between work function of metals and the saturated surface potentials. These linear relationships again suggest that the electronic charge exchange occurs at the metal/film interface. [26] The surface potential of LB films should be zero when the surface Fermi-level of films and Fermi-level of metals coincide. Therefore, judging from the value of work function at the surface potential of V = 0.0 V in Fig. 9.13, the surface Fermi level of CuttbPc LB films are estimated to be located at an electronic energy level of 4.68, 4.75 and 4.94 eV below vacuum level at temperatures of 100, 20 and -100°C, respectively. The surface Fermi-level of CuttbPc LB films shifts to a higher energy level as the temperature increases, indicating a

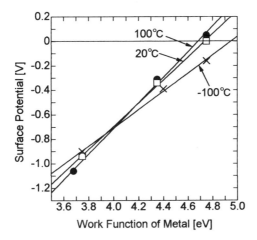

Figure 9.13 Relationship between the saturated surface potential across CuttbPc LB films and the work function of electrode materials at various temperatures.

characteristic behavior of so-called p-type semiconductors. [18, 19]

The space charge distribution due to the excess charges displaced from electrodes can be determined, [26] under the assumption that the distribution of charges is not altered by the layer by layer deposition and the relative dielectric constant of CuttbPc LB films is 2.7. Figure 9.14 shows the space charge distribution of CuttbPc LB films on Au, Cr and Al electrode at 20°C. The space charge density $\rho(x)$ steeply decreases as the number of layers n increases in the range $n < 4$. This result indicates that most of the excess charges displaced from metal electrodes exist within the first 3-4 deposited layers. The density of CuttbPc molecules is about 1.2×10^{27} m^{-3}. Thus about 0.8% of CuttbPc molecules are expected to contribute to the charge exchange at the metal/film interface. It is instructive to note here that the polarity of the space charge density $\rho(x)$ depends on metal electrodes and the number of deposited layers. For example, in the range $n = 1 - 3$, the polarity of the charge density $\rho(x)$ is positive for Au electrode, whereas it is negative for Al electrode. In other words, CuttbPc LB films have both electron donor and acceptor states at the metal/film interface, and they accept electrons from Al electrodes and donate electrons to Au electrodes. The surface Fermi level of phthalocyanine LB film and the Fermi level of metals are brought into coincident when a thermodynamics equilibrium is established at the interface. [49, 50, 51] As shown in Fig. 9.14, CuttbPc LB films on Au are positively charged at a temperature

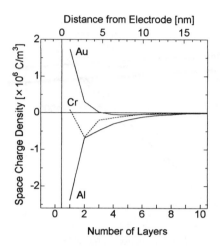

Figure 9.14 Space charge distribution in CuttbPc Langmuir-Blodgett films at a temperature of 20°C.

of 20°C, whereas the films on Al and Cr electrodes are negatively charged. This result suggests that electron donor states of CuttbPc LB films extend into in the energy level which is higher than the Fermi level of Au electrode, whereas electron acceptor states extend into the energy level which is lower than the Fermi level of Al electrodes [see Fig. 9.10(b)]. In other words, electron donor states mainly make a contribution to the built up of surface potential V_s on Au electrodes, whereas electron acceptor states mainly contribute on Cr and Al electrodes. Thus using $\rho(x)$, we obtain the estimation of

$$G(E_0, x, T) \equiv | \ \rho(x)/e \ | \cong \int_{E_0}^{+\infty} +n_D(E, x, T)dE, \tag{9.6}$$

for positively charged CuttbPc LB films on Au electrode, assuming that the Fermi-Dirac distribution function $f(E)$ is approximately given by a step function varying around the Fermi level of metals. Similarly,

$$G(E_0, x, T) \equiv | \ \rho(x)/e \ | \cong \int_{-\infty}^{E_0} +n_A(E, x, T)dE, \tag{9.7}$$

is obtained for negatively charged LB films on Al and Cr electrodes. In Eqs. (9.6) and (9.7), E_0 represents the Fermi level of metal with reference to the vacuum level, and it is given by $E_0 = -W_m + eV_s$ [see Fig. 9.9(b)]. In other words, the position

E_0 shifts due to the electrostatic potential Vs in films. Similar discussion is held for the result of surface potential of CuttbPc LB films at various temperatures, except that the electron acceptor states become dominant as the temperature decreases (see Fig. 9.13).

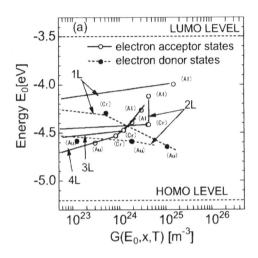

Figure 9.15 Relationship between $G(E_0, x, T)$ and the energy E_0 in CuttbPc LB film at a temperature of 20°C.

Figures 9.15 shows the $G(E_0, x, T)$ of CuttbPc LB films, respectively. Curve 1L, 2L, 3L, and 4L represent the $G(E_0, x, T)$ at positions of $x = 0.5, 1.5, 2.5$ and 3.5, respectively. Solid lines with open circles represent the $G(E_0, x, T)$ for electron acceptor states, and broken lines with closed circles for electron donor states. The marks (Al), (Cr) and (Au) indicated in Fig. 9.15 represent the $G(E_0, x, T)$ which are estimated from the data of the CuttbLB films on Al, Cr and Au shown in Fig. 9.13, respectively. It is found that very high-density electron acceptor and donor states exist at the first monolayer of CuttbPc LB films. Further, $G(E_0, x, T)$ decreases steeply as the number of deposited layers increases. From this figure, it is expected that the Fermi level of CuttbPc at 4.7 eV from the vacuum level at the interface (see curve 4L).

As described earlier, it is interesting to say that similar nanometric interfacial phenomena occur at the metal/film and metal/insulator interfaces. Further, similar results were obtained for TioPc phthalocyanine ultrathin films prepared at very slow deposition rate by organic molecular beam evaporation method. [52]

9.3 I-V Characteristic

9.3.1 PI LB Films

Excessive electronic charges transferred from metals to PI LB films exist at the metal/PI LB film interface, and the density of electronic states is very high with an order of $10^{25} - 10^{26}$ m^{-3}. As a result, very high electric field of an order of

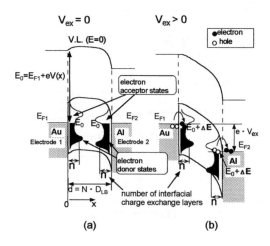

Figure 9.16 Schematic diagram of interfacial charge exchange phenomena in Au/PI LB film /Al device.

$10^8 - 10^9$ V/m is formed at the interface. Therefore it is expected that the interfacial phenomena give direct effect on the electrical transport properties of PI LB films. It is essential to gain information on the charge storage occurring at the PI/metal interface by the application of external voltage. Figure 9.16(a) shows a model of the space charge in PI LB films. The potential $V(x)$ built at position x in PI LB film due to the displaced charges is expressible as [53, 54]

$$V(x) = \frac{d-x}{d} \int_0^x \frac{x'\rho(x')}{\epsilon_0 \epsilon_s} dx' + \frac{x}{d} \int_x^d \frac{(d-x')\rho(x')}{\epsilon_0 \epsilon_s} dx' + \frac{x(E_{F2} - E_{F1} + eV_{ex})}{ed}. \quad (9.8)$$

Here, E_{F1} and E_{F2} represent the Fermi level of electrode and 2, respectively. ϵ_0 is the permittivity of a vacuum, ϵ_S is the relative dielectric constant of PI (= 3.0), [3] d is the film thickness given by $N \cdot D_{LB}$ (N: the number of deposited layers, D_{LB}: the monolayer thickness of PI LB film, and V_{ex} is the external applied voltage.

As-deposited PI LB films are charged negatively on metal electrodes, even on Au electrodes with a large work function, due to the displacement of electrons from the electrodes. [26] Electron-acceptor states broadly distribute in the range of energy levels, and they extend to the lower energy levels, which are deeper than the Fermi level of Au electrode. These states acquire excessive electrons from metal electrodes when PI LB films and metals are brought into contact. The electronic states distribute in the energy range over several hundred mili-electron volts (> thermal energy kT) in electron energy. Therefore, $f(E)$ is approximately written as a step function varying around the Fermi level, and in Eq. (9.8) is rewritten as Eq. (9.6), assuming $E_0 = E_{F1} + eV(x)$ at position x.

When external voltage V_{ex} is applied to the PI LB film, the $G(E_0, x, T)$ will be changed, possibly because the application of V_{ex} will produce an additional space charge density $\Delta\rho_1(x)$ at the electrode 1 / PI LB film interface (and $\Delta\rho_2(x)$ at the electrode 2 / PI LB film interface). Here, ΔE is given by $e\Delta V(x)$ at the electrode 1/ PI LB film interface, whereas ΔE is given by $-e(V_{ex} - \Delta V(x))$ at the electrode 2/ PI LB film interface. The potential change $\Delta V(x)$ due to the space charge density $\Delta\rho_1(x)$ and $\Delta\rho_2(x)$ is written as

$$
\begin{aligned}
dV(x) = {} & \frac{d-x}{d} \int_0^x \frac{x'[\Delta\rho_1(x') + \Delta\rho_2(x')]}{\epsilon_0\epsilon_s} dx' \\
& + \frac{x}{d} \int_x^d \frac{(d-x')[\Delta\rho_1(x') + \Delta\rho_2(x')]}{\epsilon_0\epsilon_s} dx' + \frac{xV_{ex}}{d}
\end{aligned}
\tag{9.9}
$$

with

$$
\Delta\rho_1(x) = k' \cdot e[G_1(E_0 + \Delta E, x, T) - G_1(E_0, x, T)](\equiv k' \cdot eG_1(E_0, \Delta E, x, T)) \tag{9.10}
$$

and

$$
\Delta\rho_2(x) = k' \cdot e[G_2(E_0 + \Delta E, x, T) - G_2(E_0, x, T)](\equiv k' \cdot eG_2(E_0, \Delta E, x, T)), \tag{9.11}
$$

where k' is the coefficient of electronic charge exchange ratio. $\Delta G_1(E_0, \Delta E, x, T)$ and $\Delta G_2(E_0, \Delta E, x, T)$ are the value of $G(E_0, x, T)$ change at the electrode 1/PI LB film and the electrode 2/ PI LB film, respectively. The coefficient k' is defined as unity when the interfacial electronic states acquire electrons until a quasi-thermodynamic equilibrium is established at the interface, whereas it is defined as zero when the interfacial electronic states do not accept any excessive electrons by the application of the external voltage. Since the $\Delta\rho_1(x)$ and $\Delta\rho_2(x)$ are assumed to be very small at the region close to electrodes 2 and 1, respectively, as illustrated in Fig. 9.16(b), the induced charge Q adding onto electrode 2 by the application of external voltage V_{ex} is approximately given by

$$Q = \frac{\epsilon_0\epsilon_s AV}{d} - \int_0^d \frac{x\Delta\rho_1(x)A}{d}dx + \int_0^d \frac{(d-x)\Delta\rho_2(x)A}{d}dx, \qquad (9.12)$$

where A is the electrode area of the sample. From Eq. (9.12), the capacitance C of PI LB film is calculated as

$$\begin{aligned}
C = \; & C_0[1 - \int_0^d \frac{x}{\epsilon_0\epsilon_s}\frac{\partial}{\partial V}k' \cdot e\Delta G_1(E_0, \Delta E, x, T)dx \\
& + \int_0^d \frac{(d-x)}{\epsilon_0\epsilon_s}\frac{\partial}{\partial V}k' \cdot e\Delta G_2(E_0, \Delta E, x, T)dx],
\end{aligned} \qquad (9.13)$$

with $C_0 = \epsilon_0\epsilon_s A/d$. The ratio of the capacitance change $(C - C_0)/C_0$ indicates the

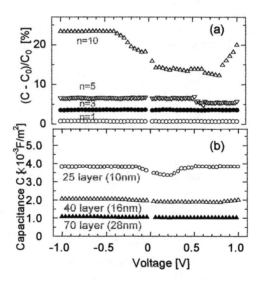

Figure 9.17 (a) C-V characteristics of Au/PI/Al device with $N = 40$, (b) C-V characteristics of Au /PI /Al device for $n = 10$.

contribution of the charges injected from electrodes.

Using the $G(E_0, x, T)$ plotted in Fig. 9.10(b), $(C - C_0)/C_0$ is calculated, assuming that the space charge distribution changes with the charge exchange ratio k' $= 1$ in the region within the first n-th layers at the metal / PI LB film interface, when samples are biased. Figure 9.17(a) shows the $(C - C_0)/C_0$ for $N = 40$. The $(C - C_0)/C_0$ increases as n increases, because the space charge injected into the films relax the external electric field V_{ex}/d. Figure 9.17(b) shows the capacitance C for $n = 10$. The capacitance increases as N decreases. The capacitance change in the range between -0.2 and 0.3 V increases as N decreases, obviously because the contribution of the charge exchange increases as N decreases. Figure 9.18 shows the relationship between

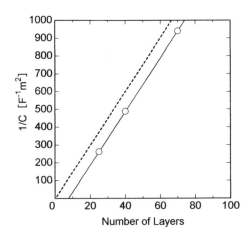

Figure 9.18 Relationship between $1/C$ and N at $n = 10$.

$1/C$ and N at $n = 3$ (curve 2) and 10, (curves 3, 4 and 5) at various electric fields. As it can be seen in the figure, a linear relationship is obtained between them. The dotted line represents the relationship between $1/C_0$ and N. The $1/C$ shifts about 8 layers to the $1/C_0$ with respect to the N-axis, that is, the apparent film thickness decreases and it is given by $(N-8) \cdot D_{LB}$ at $n = 10$ at electric fields of -2.5×10^7 V/m and $\pm 5 \times 10^7$ V/m (curve 5 in Fig. 9.18). As expected from Fig. 9.17, the decrease in the apparent film thickness depends on the electric field, and it reaches minimum at an electric field of $+2.5 \times 10^7$ V/m.

The current-voltage (I-V) characteristics of the Au/PI LB film /Al elements were measured at a temperature of 20°C by applying a triangular voltage with an amplitude V_0 of 0.5 V and a frequency f_0 of 2 mHz at a d.c. biasing voltage of -2.0, -1.0, 0.0, and 1.0 V onto Al electrode with reference to Au electrode. The capacitance of the sample was calculated using the following equation [55]

$$C = \frac{I_+ - I_-}{8 f_0 V_0}.$$ (9.14)

Here, I_+ and I_- were the currents flowing through the circuit while the applied voltage increases and decreases, respectively. The variation of the results from sample to sample was less than 20%. Figure 9.19 shows the $I - V$ and $C - V$ characteristics

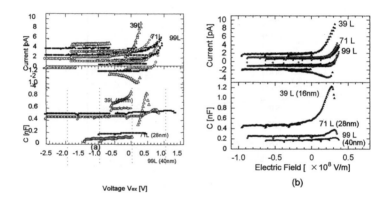

Figure 9.19 (a) $I - V$ and $C - V$ characteristics of Au / PI LB films/ Al device. (b) Electric field dependence of current and capacitance of Au/PI LB films/ Al device.

of Au/PI LB films/Al elements. The capacitance and current increase for $V_{ex} > 0$, the current is very small for $V_{ex} < 0$ and it is on the order of pico-ammeters. The current I_- decreases as the V_{ex} increases especially in the range between 0 and +0.3 V for Au/PI LB films/Al elements with 39-layer PI LB film, indicating that the discharge current increases because of the increase in the capacitance for positive biasing. Since PI has a large electron affinity, and the work function of Au is greater than that of Al, the electron injection may occur at the PI/Al interface easier than that at the PI/Au interface. In other words, it may be expected that the current flowing through the elements for $V_{ex} < 0$ is greater than that for $V_{ex} > 0$. However, the experimental results are just opposite. That is, the current flowing through the elements for $V_{ex} < 0$ is smaller than that for $V_{ex} > 0$ even when the large negative bias $(\mid V_{ex} \mid > \mid (E_{F2} - E_{F1})/e) \mid)$ is applied to the film. These results will be explained by assuming that the very high electric field with an order of 10^9 V/m gives a significant contribution for the electron injection. That is, the electron injection is restricted at the PI/Al interface, whereas it is not at the Au/PI interface. Figure 9.18(b) shows the external electric field E $(= V_{ex}/d)$ dependence of current and capacitance C of the Au/ PI LB films/ Al elements. Of interest is that the initial rise of the current and capacitance appears at the same electric field for 39, 71, 99 layers.

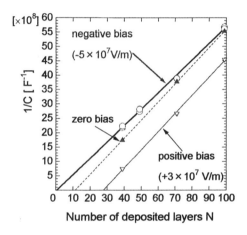

Figure 9.20 Relationship between $1/C$ and N and its electric field dependence of Au/PI LB film/Al device.

Figure 9.20 shows the relationship between $1/C$ and the number of deposited layers. A linear relationship is observed. The relative dielectric constant estimated from the slope of the linear relationship is nearly 3. The apparent film thickness decrease is seen in Fig. 9.20, in a manner similar to that seen in Fig. 9.18. For negative biasing, the apparent film thickness decrease is very small, and the charge exchange mentioned earlier occurred in the range within the first 3 LB layer at the PI/Al interface, probably because the very high electric field ($\sim 10^9$ V/m) in the direction opposite to the external applied field prevented the charge injection at the PI/Al interface. In contrast, for positive biasing, the apparent film thickness decrease is much larger than that for $V_{ex} < 0$ and the film thickness decreases monotonously as the external electric field increases. The charge exchange easily occurred at the PI/Al interface, possibly because the electric field formed by the space charge layer at this interface worked to assist the charge injection. [56, 57, 58]

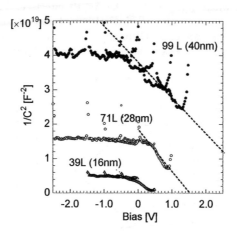

Figure 9.21 $1/C^2 - V$ characteristics of Au/PI LB film /Al device.

Figure 9.21 shows the $1/C^2 - V$ characteristic of Au/PI LB film /Al element obtained by re-plotting the C-V curves in Fig. 9.19(a). The magnitude of $1/C^2$ decreases linearly as V_{ex} increases for $V_{ex} > 0$, indicating that the apparent film thickness of insulating layer in Au/PI LB film/ Al device decreases possibly due to the charge injection from PI/Al interface. In other words, the results seen in Fig. 9.21 just correspond to the apparent film thickness decrease seen in Fig. 9.20 for positive biasing, and not to the so-called Schottky- layer thickness change observed in metal/semiconductor junctions by the application of biasing voltage. [18, 19] As shown in Figs. 9.19-9.21, PI LB film has a very high resistivity ($> 10^{14}$ $\Omega \cdot$m) even under a very high electric field of 10^8 V/m ($V_{ex} < 0$). Thus, PI LB film is a promising material working as a good thin insulating layer in the field of micro and/or molecular electronics. However the interfacial electrostatic phenomena at the metal/film interface should be taken into consideration before use.

9.3.2 Semiconductor Films

As mentioned earlier, CuttbPc LB films are charged due to the excess charges displaced from metal electrodes. Thus it is expected that the Au/CuttbPc/Al elements are charged in a manner as shown in Fig. 9.16(a), when Au and Al electrodes are electrically shorted, because a negatively charged layer is formed at the Al/CuttbPc LB film interface and a positively charged layer is formed at the Au/CuttbPc LB film interface. The thickness of the negatively charged layer and of positively charged layer are estimated to be about 10 nm and 3 nm from Figs. 9.12 and 9.14. The space charge density $\rho(x)$ at the position x from Al electrode is expressed by Eq. (9.1). On the one hand, when an external voltage V_{ex} is applied to the Au electrode with reference to the Al electrode, the thickness of space charge layer will be changed as shown in Figs. 22(a) and (b), because the application of V_{ex} will produce the addi-

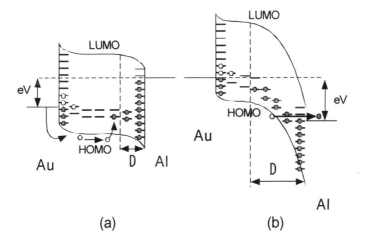

(a)   (b)

Figure 9.22 Schematic illustration of the energy structure of the Au/CuttbPc LB film/Al element.

tional space charge density $\Delta\rho(D_s)$ at the metal/ CuttbPc LB film interface. The relationship between the applied voltage and the space charge density $\rho(x)$ at the Al/CuttbPc LB film interface is given by Eq. (9.15) under the following assumptions: (i) redistribution of charges in films in the region of $0 < x < D_s$ is negligibly small, and (ii) the applied voltage V_{ex} is added to the space charge layer at the Al/CuttbPc LB film interface. [59]

$$V_s + V_{ex} = \int_0^{D_s+\Delta D} \frac{x\rho(x)}{\epsilon_0 \epsilon_r} dx. \tag{9.15}$$

Here, $\rho(x)$ is the space charge density at the position x from Al upper electrode. V_s is the electrostatic potential built across Al/CuttbPc LB film interface when $V_{ex} = 0$. The thickness of the space charge layer increases for $V_{ex} < 0$, whereas it decreases for

$V_{ex} > 0$. Differentiating Eq. (9.15) with respect to V_{ex}, the relationship between the space charge density $\rho(D_s)$ and the capacitance C is obtained, which is given by

$$\rho(D_s) = \frac{2}{\epsilon_0 \epsilon_r} \frac{\Delta V_{ex}}{\Delta(1/C^2)}, \qquad (9.16)$$

assuming that the negatively charged Al/CuttbPc LB film layer functions as an insulating layer, and the capacitance per unit area of Au/CuttbPc LB film/ Al element is expressed by $C = \epsilon_0 \epsilon_r / D_s$.

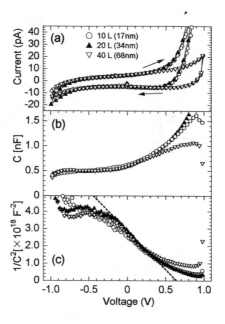

Figure 9.23 Electrical characteristics of Au/CuttbPc LB film/Al elements. (a) $I - V$, (b) $C - V$, and (c) $1/C^2 - V$ characteristics.

Figure 9.23 shows the $I-V$, $C-V$ and $1/C^2-V$ characteristics of the Au/CuttbPc LB film/Al element at a frequency of 2 mHz. The I-V characteristic shows a rectifying behavior similar to a Schottky-type diode, [18, 19] and the current I increases steeply for $V_{ex} > 0$. Further, the capacitance increases as the applied voltage V_{ex} increases in the range $-0.5 < V_{ex} < 1.0$ V. The $I - V$ and $C - V$ characteristics do not depend on the film thickness in the range $-1.0 < V_{ex} < +0.5$ V. From the electrical measurement, it is found that the Al/CuttbPc LB film interface has a very high electrical resistance of an order of 10^{14} Ω·cm, whereas the bulk and/or Au/CuttbPc LB film interface has a relatively small resistance of an order of 10^8 Ω·cm. Therefore it is expected that the applied field V_{ex} is added to the Al/CuttbPc interfacial layer, and that the Al/CuttbPc LB film interfacial layer well blocks current flowing across the film in the range $-1.0 < V_{ex} < +0.5$ V. From Eq. (9.15) and Fig. 9.23(b), the thickness of insulating layer is estimated to be about 11 nm at zero bias, and this film thickness just corresponds to the thickness of negatively charged layer at the Al/CuttbPc LB film interface. The thickness of Al/CuttbPc insulating layer D_s decreases as V_{ex} increases, where D_s is 5 nm in the maximum. The capacitance and current increase gently in the range $V_{ex} < -0.5$ V. As mentioned in section 9.2.1, the Fermi level of CuttbPc LB films locates at an energy of 4.7 eV, whereas the highest occupied molecular orbital (HOMO) level of CuttbPc LB films is estimated to locate at an energy of 5.2 eV from the cyclic voltammetry measurement (not shown here). Thus for $V_{ex} < -0.5$ V, the HOMO levels becomes higher than the Fermi level of Al electrode, and a very high electric field with an order of 10^6 V/cm is applied to the Al/CuttbPc LB film interface. Therefore at the Al/CuttbPc LB film interface, the current is allowed to flow in the range $V_{ex} < -0.5$ V, probably due to the electron tunneling or hole injection between HOMO level of CuttbPc LB film and Al electrode (see Fig. 9.22(b)).

In Fig. 9.23(c), the $1/C^2$ increases gradually as V_{ex} decreases in the range for $-0.2 < V_{ex} < +0.4$ V. A slope of the $1/C^2 - V$ characteristic gives the density of electron acceptor states, and the density is calculated using Eq. (9.16) and Fig. 9.23(c). Figure 9.24 shows the results obtained. The distribution of space charge density obtained using the surface potential measurement is also plotted by solid line. It is interesting to note that the distribution of $\rho(D_s)$ obtained by capacitance measurement agrees with that obtained by surface potential measurement for $3 < D_s < 10$ nm, suggesting that most of excessive electronic charges in this region can move by the application of the external voltage with a frequency of an order of mHz. On the contrary, the values of $\rho(D_s)$ are scattered in the range $D_s \sim 3$ nm and the distribution of $\rho(D_s)$ can not be estimated. It is interesting to note here that the thickness of $D_s \sim 3$ nm corresponds to the thickness of the first 2-molecular layer. Thus, from the viewpoint of the determination of the space distribution, it is concluded that the surface potential method is more helpful to determine the distribution in the range within several nanometers from the interface. However, from the viewpoint of device operation, the information on the charge exchange at the interface is required and the $C - V$ measurement is helpful. Therefore, the combination of the surface potential method and $C - V$ measurement will be important for further development of the organic molecular devices, and for a complete understanding of the electronic phenomena at interfaces.

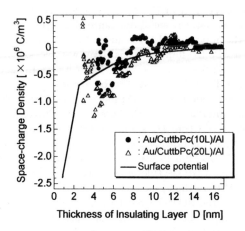

Figure 9.24 Space charge distribution of the Al/CuttbPc LB film interface determined from the $C - V$ measurement and the surface-potential measurement.

9.4 Summary

In this chapter, the importance of the concept of the monolayer/material interfaces for the development of the theory of monolayer dielectrics and for the application of monolayers in the field of electronics has been revealed. Further, it has also been shown that successfully prepared ultra-thin dielectric films can be used such as a tunneling spacer in electronic devices. Finally, it has been suggested that the understanding of the electronic phenomena occurring at the interface will be key for the development of organic materials devices including monolayer film devices.

References

[1] M. Iwamoto and S. Mashiko, *Proceedings of the Nano-molecular Electronics International Workshop 97'*, Thin Solid Films, **331** (1998).

[2] G. G. Roberts, *Langmuir-Blodgett Films*, Plenum Press, NewYork, (1990) Chap. 7.

[3] M. Iwamoto and M. Kakimoto, *Polyimides as Langmuir-Blodgett Films*, in *Polyimides, Fundamentals and Applications*, ed. Malay K. Ghosh and K. L. Mittal, Marcel Dekker, Inc., New York (1996) Chap. 25.

[4] M. Iwamoto, T. Kubota, M. Nakagawa, and M. Sekine, *J. Phys. D: Appl. Phys.*, **23** (1990) 575.

[5] T. Kubota, M. Iwamoto, H. Noshiro, and M. Sekine, *Jpn. J. Appl. Phys.*, **L30** (1991) 393.

[6] W. Schutt, H. Koster, and G. Zuther, *Thin Solid Films*, **31** (1976) 275.

[7] S. Carraral, *Nano-particles and Nanostructured Films*, ed. J. H. Fendl, Wiley-VCH, Weinheim (1998) Chap. 15.

[8] M. Sugi, *Structure-dependent Carrier Transport in Langmuir Multilayer Assembly Films*, Researches of the Electrotechnical Laboratory, No. 794 (1978).

[9] G. J. Ashwell, *Molecular Electronics*, Research Studies Press, New York (1992).

[10] C. M. Fischer, M. Burghard, S. Roth and K. V. Klitzing, *Europhys. Lett.*, **28** (1994) 12.

[11] C. M. Fischer, M. Burghard, S. Roth and K. V. Klitzing, *Appl. Phys. Lett.*, **26** (1995) 3331.

[12] M. Iwamoto, M. Wada, and T. Kubota, *Thin Solid Films*, **244** (1994) 472.

[13] Y. Noguchi, Y. Majima, M. Iwamoto, T. Kubota, T. Nakahama, S. Yokoyama and S. Mashiko, IEICE Trans. (2000) in press.

[14] Special Issue, *Functional Organic Materials for Devices*, J. Material. Chem., **9** (1999) 1853-2276.

[15] Special Issue, *Nano-Molecular Electronics*, Jpn. J. Appl. Phys., **34** (1995).

[16] A.Ulman, *Ultrathin Organic Thin Films*, Academic Press, San Diego (1991).

[17] T. Nakamura, in *Handbook of Organic Conductive Molecules and Polymers.*, **Vol.1**, *Charge transfer Salts, Fllerenes and photoconductors*, ed. H. S. Nalwa, John Wiley & Sons Ltd., New York (1997) Chap. 14.

[18] S. M. Sze, *Physics of Semiconductor Devices*, John and Wiley & Sons, New York (1981).

[19] H. Meier, *Organic Semiconductors, Dark-and Photo-conductivity of Organic Solids*, Verlag Chemie, in *Monographs in Modern Chemistry* (1974).

[20] L. S. Hung, C. W. Tang, and M. G. Mason, *Appl. Phys. Lett.*, **70** (1997) 152.

[21] A. A. Sonin, *The Surface Physics of Liquid Crystals*, Gordon Breach Publishers, Amsterdam (1995).

[22] P. K. Watson, *IEEE Trans. Dielec. & Elect. Insul.*, **2** (1995) 915, and references cited therein.

[23] T. J. Lewis, *IEEE. Trans. Dielect. & Elect. Insul.*, **1** (1994) 812.

[24] J. Lowell and A. C. Rose-Innes, *Advances in Physics*, **29** (1980) 947, and references cited therein.

[25] L. H. Lee, *J. Electrostatics*, **32** (1994) 1, and references cited therein.

[26] M. Iwamoto, A. Fukuda, and E. Itoh, *J. Appl. Phys.*, **75** (1994) 1607.

[27] M. C. Petty, *Langmuir-Blodgett films*, Cambridge, New York (1996).

[28] D. S. Soane and Z. Martynenko, *Polyimide in Microelectronics - Fundamentals and Applications-*, Elsevier, Amsterdam (1989).

[29] E. Sugimoto, *IEEE. Trans. Electr. Insul. Mag.*, **5** (1989) 15.

[30] F. H. Moser and A. L. Thomas, *The Phthalocyanines*, CRC Press, Boca Raton, FL (1983).

[31] C. C. Leznoff and A. B. P. Lever, *Phthalocyanines-Properties and Applications*, VCN Publisheres, Weinheim (1989).

[32] S. Baker, M. C. Petty, G. G. Roberts, and M. V. Twigg, *Thin Solid Films*, **99** (1983) 53.

[33] G. M. Sessler, *IEEE Trans. Dielectrics and Electr. Insul.*, **4** (1997) 614, and references cited therein.

[34] N. H. Ahmed and N. N. Srinvas, *IEEE Trans. Dielectrics and Electr. Insul.*, **4** (1997) 644, and references cited therein.

[35] R. E. Collins, *Appl. Phys. Lett.*, **26** (1975) 675.

[36] G. M. Sessler, J. E. West, and G. Gerhard, *Phys. Rev. Lett.*, **48** (1982) 563.

[37] G. M. Sessler, J. E. West, and D. A. Berkley, *Phys. Rev. Lett.*, **38** (1977) 368.

[38] T. Takada, T. Maeno and H. Kusibe, *IEEE Trans. Electr. Insul. EI-22* (1987) 497.

[39] E. Itoh and M. Iwamoto, *J. Appl. Phys.*, **81** (1997) 1790.

[40] M. Iwamoto and E. Itoh, *Thin Solid Films*, **331** (1998) 15, and references cited therein.

[41] R. H. Tredgold and G. W. Smith, *Thin Solid Films*, **99** (1983) 215.

[42] R. H. Tredgold and G. W. Smith, *J. Phys. D: Appl. Phys.*, **14** (1981) L193.

[43] E. Itoh and M. Iwamoto, *J. Electrostat.*, **36** (1996) 313.

[44] G. M. Sessler, *J. Appl. Phys.*, **43** (1972) 405.

[45] H. Kirihata and M. Uda, *Rev. Sci. Instrumn.*, **52** (1981) 68.

[46] E. Itoh and M. Iwamoto, *Appl. Phys. Lett.*, **68** (1996) 2714.

[47] C. B. Duke and T. J. Fabish, *Phys. Rev. Lett.*, **37** (1976) 1075.

[48] T. J. Fabish and C. B. Duke, *J. Appl. Phys.*, **75** (1974) 1607.

[49] E. Itoh, H. Kokubo, S. Shouriki, and M. Iwamoto, *J. Appl. Phys.*, **83** (1998) 372.

[50] H. Kokubo, Y. Oyama, Y. Majima, and M. Iwamoto, *J. Appl. Phys.*, **86** (1999) 3848.

[51] M. Iwamoto, *J. Mater. Chem.*, **10** (2000) 99, and references cited therein.

[52] Y. Majima, K. Yamagata, and M. Iwamoto, *J. Appl. Phys.*, **86** (1999) 3229.

[53] E. Itoh and M. Iwamoto, *J. Appl. Phys.*, **85** (1999) 7239.

[54] C. Q. Li, C. X. Wu, and M. Iwamoto, *Jpn. J. Appl. Phys.*, **39** (2000) 1840.

[55] A. J. Twarowski and A. C. Albrecht, *J. Chem. Phys.*, **70** (1979) 2255.

[56] G. M. Sessler, B. Hahn, and D. Y. Yoon, *J. Appl. Phys.*, **60** (1986) 318.

[57] Z. L. Wu and G. R. Govinda Raju, *IEEE. Trans. Dielectr. Insul.*, **2** (1995) 475.

[58] Y. Suzuki, H. Muto, T. Mizutani and M. Ieda, *J. Phys. D; Appl. Phys.*, **18** (1985) 2293.

[59] E. Itoh and M. Iwamoto, *Jpn. J. Appl. Phys.*, **38** (1999) 5945.

INDEX

Atomic Force Microscopy (AFM), 1, 11
Azobenzene, 61, 62, 69, 70

Bessel function, 82
Boltzmann constant, 22, 85, 146, 159
Boltzmann statistic, 57, 82, 85
Born-Oppenheimer (BO), 11
Bragg-Williams approximation, 119
Bragg-Williams cholesterics, 126
Bragg-Williams Mixing Energy, 118, 137
Bragg-Williams theory, 131-132, 142

C_∞ symmetry, 143
$C_{\infty v}$ symmetry, 26, 32, 78, 86, 144
CD effect, 129
chiral discrimination (CD), 10, 141
charality, 58, 59, 60
chirality representation, 144
chiral molecules, 12
chiral phase separation (CPS), 10, 11, 117, 120, 124, 125, 126, 130, 131
Chiral Phospholipid , 57
chiral symmetry breaking (CSB), 10, 11, 117
Chiral terms, 144
Closed Packing, 105
compression speed, 39
concentration square gradient (CSG), 127
conduction current, 38
coulomb staircase, 170
critical molecular area, 89, 91, 97
Cu-tetra-(tertbutyl)-phthalocyanine (CuttbPc), 14
Cyclohexanecarboxylate-type liquid crystal (DON), 45, 54

debye Brownian Motion Model, 96, 107
Debye-Langevin equation, 22
Dielectric Anisotropy, 25, 40, 157
Dielectric Constant, 23, 29, 33, 89